A-Level Year 2

Physics

Exam Board: OCR B (Advancing Physics)

Revising for Physics exams is stressful, that's for sure — even just getting your notes sorted out can leave you needing a lie down. But help is at hand...

This brilliant CGP book explains **everything you'll need to learn** (and nothing you won't), all in a straightforward style that's easy to get your head around. We've also included **exam questions** to test how ready you are for the real thing.

There's even a free Online Edition you can read on your computer or tablet!

How to get your free Online Edition

Go to **cgpbooks.co.uk/extras** and enter this code...

1480 2683 6430 7432

This code only works for one person. If somebody else has used this book before you, they might have already claimed the Online Edition.

A-Level revision? It has to be CGP!

Published by CGP

Editors:
Emily Garrett, David Maliphant, Rachael Marshall, Sam Pilgrim, Frances Rooney, Charlotte Whiteley, Sarah Williams and Jonathan Wray.

Contributors:
Stuart Barker, Peter Cecil, Mark A. Edwards, Barbara Mascetti, John Myers, Zoe Nye, Moira Steven and Tony Winzor.

ISBN: 978 1 78294 345 7

With thanks to Jan Greenway for the copyright research.

Clipart from Corel®
Printed by Elanders Ltd, Newcastle upon Tyne.

Based on the classic CGP style created by Richard Parsons.

Contents

The Scientific Process

'How Science Works' is all about the scientific process — how we develop and test scientific ideas. It's what scientists do all day, every day (well, except at coffee time — never come between a scientist and their coffee).

Scientists Come Up with Theories — Then Test Them...

Science tries to explain **how** and **why** things happen — it **answers questions**. It's all about seeking and gaining **knowledge** about the world around us. Scientists do this by **asking** questions, **suggesting** answers and then **testing** their suggestions to see if they're correct — this is the **scientific process**.

1) **Ask** a question about **why** something happens or **how** something works. E.g. why are distant galaxies moving away from us?
2) **Suggest** an answer, or part of an answer, by forming a **theory** (a possible **explanation** of the observations) — e.g. because the universe began with a hot big bang, and has been expanding ever since. (Scientists also sometimes form a **model** too — a **simplified picture** of what's physically going on.)
3) Make a **prediction** or **hypothesis** — a **specific testable statement**, based on the theory, about what will happen in a test situation. For example, if the universe began with a hot big bang, we should still be able to see the electromagnetic radiation produced in the early stages of the universe.
4) Carry out a **test** — to provide **evidence** that will support the prediction (or help to disprove it). E.g. sending satellites to investigate the cosmic microwave background radiation (p.33).

According to James and Archie's theory of cyclical fashion, it was only a matter of time before their look was back on top.

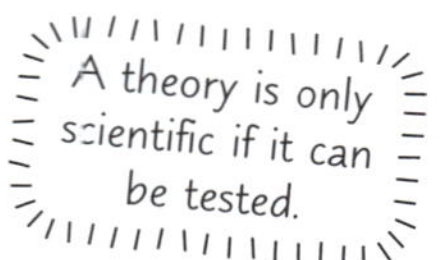

...Then They Tell Everyone About Their Results...

The results are **published** — scientists need to let others know about their work. Scientists publish their results in **scientific journals**. These are just like normal magazines, only they contain **scientific reports** (called papers) instead of the latest celebrity gossip.

1) Scientific reports are similar to the **lab write-ups** you do in school. And just as a lab write-up is **reviewed** (marked) by your teacher, reports in scientific journals undergo **peer review** before they're published.
2) The report is sent out to **peers** — other scientists that are experts in the **same area**. They examine the data and results, and if they think that the conclusion is reasonable it's **published**. This makes sure that work published in scientific journals is of a **good standard**.
3) But peer review **can't guarantee** the science is **correct** — other scientists still need to **reproduce** it.
4) Sometimes **mistakes** are made and bad work is published. Peer review **isn't perfect** but it's probably the best way for scientists to self-regulate their work and to publish **quality reports**.

...Then Other Scientists Will Test the Theory Too

Other scientists read the published theories and results, and try to **test the theory** themselves. This involves:

- Repeating the **exact same experiments**.
- Using the theory to make **new predictions** and then testing them with **new experiments**.

If the Evidence Supports a Theory, It's Accepted — for Now

1) If all the experiments in all the world provide good evidence to back it up, the theory is thought of as **scientific 'fact'** (for now).
2) But it will never become **totally indisputable** fact. Scientific **breakthroughs or advances** could provide new ways to question and test the theory, which could lead to **new evidence** that **conflicts** with the current evidence. Then the testing starts all over again...

And this, my friend, is the **tentative nature of scientific knowledge** — it's always **changing** and **evolving**.

The Scientific Process

So scientists need evidence to back up their theories. They get it by carrying out experiments, and when that's not possible they carry out studies. But why bother with science at all? We want to know as much as possible so we can use it to try and improve our lives (and because we're nosy).

Evidence Comes From Controlled Lab Experiments...

1) Results from **controlled experiments** in **laboratories** are **great**.
2) A lab is the easiest place to **control variables** so that they're all **kept constant** (except for the one you're investigating).

...That You can Draw Meaningful Conclusions From

1) You always need to make your experiments as **controlled** as possible so you can be confident that any effects you see are linked to the variable you're changing.
2) If you do find a relationship, you need to be careful what you conclude. You need to decide whether the effect you're seeing is **caused** by changing a variable (this is known as a **causal relationship**), or whether the two are just **correlated**.

Using all the available evidence, Andy had concluded that staying on the beach too long turned you into a seal.

Society Makes Decisions Based on Scientific Evidence

1) Lots of scientific work eventually leads to **important discoveries** or breakthroughs that could **benefit humankind**.
2) These results are **used by society** (that's you, me and everyone else) to **make decisions** — about the way we live, what we eat, what we drive, etc.
3) All sections of society use scientific evidence to make decisions, e.g. politicians use it to devise policies and individuals use science to make decisions about their own lives.

Other factors can **influence** decisions about science or the way science is used:

Economic factors

Society has to consider the **cost** of implementing changes based on scientific conclusions — e.g. the cost of reducing the UK's carbon emissions to limit the human contribution to **global warming**.

Scientific research is often **expensive**. E.g. in areas such as astronomy, the Government has to **justify** spending money on a new telescope rather than pumping money into, say, the **NHS** or **schools**.

Social factors

Decisions affect **people's lives** — e.g. when looking for a site to build a **nuclear power station**, you need to consider how it would affect the lives of the people in the **surrounding area**.

Environmental factors

Many scientists suggest that building **wind farms** would be a **cheap** and **environmentally friendly** way to generate electricity in the future. But some people think that because **wind turbines** can **harm wildlife** such as birds and bats, other methods of generating electricity should be used.

So there you have it — how science works...

Hopefully these pages have given you a nice intro to how science works, e.g. what scientists do to provide you with 'facts'. You need to understand this, as you're expected to know how science works yourself — for the exam and for life.

Radioactivity and Exponential Decay

Making models in physics is nothing to do with squeezy bottles and sticky-back plastic, I'm afraid. But models are designed to make life easier — they simplify things and show links between topics you might not think are connected.

Models Use Assumptions to Simplify Problems

1) A **model** is a **set of assumptions** that **simplifies** and **idealises** a particular situation.
2) The assumptions mean you can write **equations** that describe what's going on, allowing you to make **calculations** and **predictions**. Without the assumptions, there would be **too many factors** to consider.
3) The topics in this section may appear to be **unconnected**, but they're all based on **models** that use **differential equations** (see below) to describe the **rate of change** of something.
4) This is another **benefit** of models — once you have a model that describes one process you can often **extend it**, or **change it slightly**, to describe **another**, **unrelated process** without having to start all over again — compare the models for **radioactive decay** (p.6) and **capacitor discharge** (p.9) and you'll see what I mean.

Unstable Nuclei are Radioactive

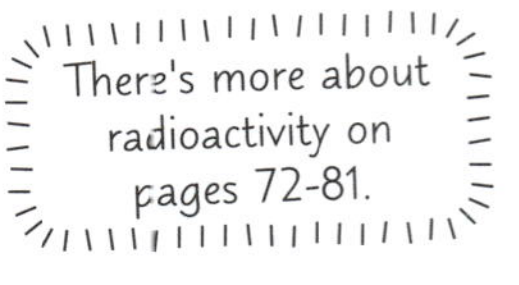

1) If a nucleus has **too many neutrons**, **not enough neutrons**, **too many nucleons** in total, or **too much energy**, it may be **unstable**.
2) Unstable nuclei **break down** by **releasing energy** and/or **particles**, until they reach a **stable form** — this process is called **radioactive decay**.
3) Radioactive decay is a **random** process — you **can't** tell when any **one nucleus** will **decay**, or **which nucleus** in a sample will be the **next** to decay.

This radio was pretty active...

Radioactivity can be Modelled by Exponential Decay

1) Although you can't predict when a nucleus will decay, you can still make predictions about the behaviour of a radioactive source using a **model** — **exponential decay** — based on a **very large number** of undecayed nuclei.
2) If you take a **large** enough **sample** of **unstable nuclei**, the **overall behaviour** shows a **pattern**. This means you can predict **how many nuclei** will decay in a **given time** (even though you can't predict which ones).
3) If you plot a graph showing the **number** of unstable nuclei that **decay each second** against **time**, you always get an **exponential decay curve**, like the one on page 5 — so this can be used as a **model** for radioactive decay.

The Rate of Decay is Measured by the Decay Constant

The **number** of unstable nuclei that **decay each second** is called the **activity** of the sample. The **activity** of a sample is **proportional** to its size — as **nuclei decay**, the **sample** size gets **smaller**, so the **activity falls**. If you plot a graph of **activity** against time, you'll see that the **gradient** gets **shallower and shallower**. This kind of graph, where the rate of change decreases as the quantity gets smaller, is called an **exponential decay** curve.

The **decay constant** (λ) measures how **quickly** an isotope will **decay** — it's the **probability** of a given nucleus **decaying** in a certain **time**. The **bigger** the value of λ, the **more likely** a decay is, so the **faster** the rate of decay. Its unit is s^{-1}. For a given isotope, λ is a **fixed** probability (it doesn't change at all).

activity = decay constant × number of undecayed nuclei In symbols: $A = \lambda N$

Don't get λ confused with wavelength.

Activity is measured in **becquerels** (Bq). An activity of 1 Bq means that 1 nucleus decays every second.

If you plot a graph of the **number of undecayed nuclei** remaining in a sample (N) against time (see next page), the **gradient** of the graph is **negative**. The gradient is the change in the number of radioactive nuclei remaining in a given time (the rate of decay), which must also be **negative**.

$$\frac{dN}{dt} = -\lambda N$$

Equations like $\frac{dN}{dt} = -\lambda N$ are a type of **differential equation**. They're used to describe the **rate of change** of something, and are really helpful for modelling situations where the rate of change of a quantity **is proportional to the value** of that quantity (like exponential decay). In the case of radioactive decay, the rate of decay is proportional to the number of undecayed nuclei remaining. Equations like this have the general form: $\frac{dx}{dt} = -kx$, where k is a constant.

You can model the charge remaining on a discharging capacitor in the same way (p.9).

Radioactivity and Exponential Decay

*You can **Find** the **Half-Life** of an **Isotope** from a **Graph***

The **half-life** ($T_{1/2}$) of an **isotope** is the **average time** it takes for the **number of undecayed nuclei** to **halve**.

The **longer** the **half-life** of an isotope, the **longer** a sample of a given initial size will remain **radioactive**.
You can find the **half-life** of an isotope from a graph of the number of **undecayed nuclei** remaining against **time**:

1) Read off the original number of undecayed nuclei (when $t = 0$), N_0.
2) Go to **half** the original value of N.
3) Draw a **horizontal line** to the **curve**, then a **vertical line** down to the ***x*-axis**.
4) The **half-life** is where this line meets the *x*-axis.
5) It's always a good idea to **check** your answer. Repeat steps 1-3 for a **quarter** of the original value of N, and divide the time where the line meets the *x*-axis by **two**. That will also give you the half-life. You can do the same for an eighth of the original value (divide the time by 3), and a sixteenth of the original value (divide the time by 4). Check that you get the **same answer** each time.

If you're not sure what an isotope is, see p.72.

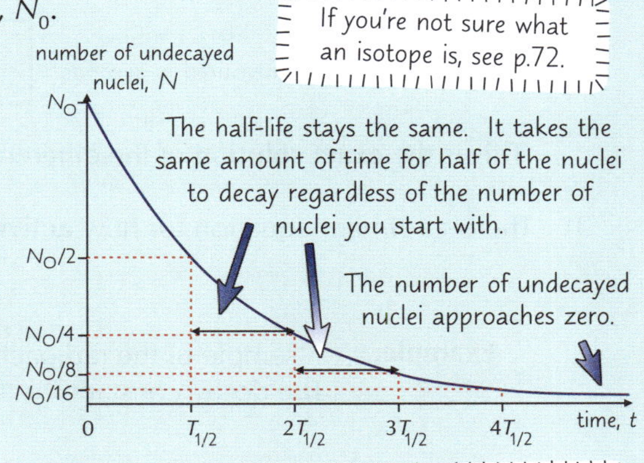

In practice, half-life isn't measured by counting undecayed nuclei (which can be very tricky), but by measuring the **time it takes** for the **activity** (or the count rate) to **halve**. The method is **exactly the same**, you just plot activity (or count rate) against time rather than the number of undecayed nuclei.

Activity is the number of decays per second. Count rate is the number of decays detected per second. As detectors generally only detect radiation that's emitted in one direction, but sources emit radiation in all directions, the count rate will be proportional to, but less than, the source's activity.

*You can Generate a **Count-Rate** Decay Graph **Experimentally***

You're most likely to do this using the isotope **protactinium-234**.
Protactinium-234 is formed when **uranium** decays (via another isotope).
You can measure protactinium-234's decay rate using a **protactinium generator** — a bottle containing a uranium salt, the decay products of uranium (including protactinium-234) and two solvents, which separate out into layers, like this:

layer containing protactinium-234

layer containing uranium salt

1) **Shake the bottle** to mix the solvents together, then add it to the equipment shown on the right.
2) Wait for the liquids to **separate**. The protactinium-234 will be in solution in the top layer, and the uranium salt will stay in the bottom layer. Then you can point the Geiger-Müller tube at the top layer to measure the activity of the protactinium-234.
3) As soon as the liquids separate, record the count rate. Re-measure the count rate at sensible intervals (e.g. every 30 seconds).
4) Once you've collected your data, leave the bottle to stand for at least ten minutes (this should be long enough for all the protactinium in the top layer to decay), then take the count rate again. This is the **background count rate** corresponding to background radiation — the low level radiation that you get everywhere. (You could also do this at the beginning of the experiment, before shaking the bottle).
5) **Subtract this value** from your measured count rates, then plot a graph of count rate against time. It should look like the graph on the right. You can use this graph to find the **half-life** in exactly the same way as above. In this case the half-life is the time taken for the count rate to halve.

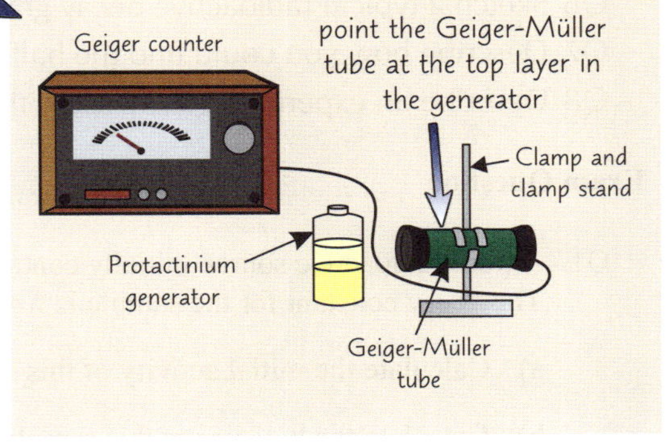

count rate

time, t

You could also connect the Geiger-Müller tube to a data logger and a computer to record this data automatically, then use the computer to generate the graph (you'll still need to remove the background count rate though).

Radioactivity and Exponential Decay

You Need to Know the Equations for Half-Life and Decay

1) The **half-life** can be **calculated** using the equation: $T_{1/2} = \frac{\ln 2}{\lambda}$ where $T_{1/2}$ is the half-life (in s), ln is the natural log (see page 84) and λ is the decay constant (in s^{-1})

2) The **number of undecayed nuclei** remaining, N, depends on the **number originally** present, N_0. The **number remaining** can be calculated using the equation:

Here t = time, measured in seconds. $N = N_0 e^{-\lambda t}$

This is the equation of the curve on the graph at the top of page 5.

This is the **exact solution** of the differential equation $\frac{dN}{dt} = -\lambda N$.

3) There's a similar equation for how **activity** changes with time: $A = A_0 e^{-\lambda t}$ where A is the activity (in Bq) and A_0 is the initial activity

Example: A sample of the radioactive isotope ^{13}N contains 5.0×10^6 undecayed nuclei. The decay constant for this isotope is $1.16 \times 10^{-3}\ s^{-1}$.

a) What is the half-life for this isotope?

$T_{1/2} = \frac{\ln 2}{\lambda} = \frac{\ln 2}{1.16 \times 10^{-3}} = 597.54... =$ **598 s (to 3 s.f.)**

b) How many undecayed ^{13}N nuclei will remain after 810 seconds?

$N = N_0 e^{-\lambda t} = (5.0 \times 10^6) \times e^{-(1.16 \times 10^{-3} \times 810)} = 1.953... \times 10^6$
= 2.0×10^6 nuclei (to 2 s.f.)

Practice Questions

Q1 What is meant by the statement 'radioactive decay is a random process'?

Q2 What does the decay constant of a radioactive isotope correspond to?

Q3 Define radioactive activity. What units is it measured in?

Q4 What is the general form of an equation where the rate of change of a quantity is proportional to the value of that quantity?

Q5 What is meant by the term 'half-life'?

Q6 Sketch a typical radioactive decay graph showing the number of undecayed nuclei against time.

Q7 Describe how you could find the half-life of a radioactive isotope from a graph of activity against time.

Q8 Describe an experiment to measure the half-life of protactinium-234.

Exam Question

Q1 A pure radioactive source initially contains 51 000 undecayed nuclei. The decay constant for the sample is $\lambda = 0.014\ s^{-1}$.

a) Calculate the initial activity of this sample. [1 mark]

b) Calculate the half-life of this sample. [1 mark]

c) Estimate how many nuclei of the radioactive source there will be after 300 seconds. [1 mark]

d) Exponential decay is used to model radioactive decay. Give one reason why models are useful in science. [1 mark]

Radioactivity is a random process — just like revision shouldn't be...

Remember the shape of that graph — whether it's count rate, activity or number of nuclei plotted against time, the shape's always the same. The maths is a bit of a pain, but I think the experiment's pretty good.

Capacitors

You might not have guessed that a page on radioactive decay would be followed by a page on capacitors, but by the time you get to pages 8-11 it'll make perfect sense. Anyway, first you need to learn about capacitors...

Capacitance is Defined as the Amount of Charge Stored per Volt

Capacitors are made up of two metal plates separated by an air gap or an insulator (so no charge can flow between them). These plates **store electrical charge** — a bit like a charge bucket. **Capacitance, *C***, is a measure of **how much charge** a capacitor can hold — it's defined as the **amount of charge stored per volt**.

$$C = \frac{Q}{V}$$

Q is the charge in coulombs and *V* is the potential difference in volts

Capacitance is measured in **farads** (F). **1 farad = 1 C V⁻¹** (i.e. one coulomb per volt).

A farad is a **huge** unit, so you'll usually see capacitances expressed in terms of:

μF — microfarads ($\times 10^{-6}$), **nF** — nanofarads ($\times 10^{-9}$), **pF** — picofarads ($\times 10^{-12}$)

Remember the link between **current, *I*, charge, *Q***, and **time, *t***, too — you'll often need **both equations** for capacitor questions.

$$I = \frac{\Delta Q}{\Delta t}$$

Capacitors Also Store Energy

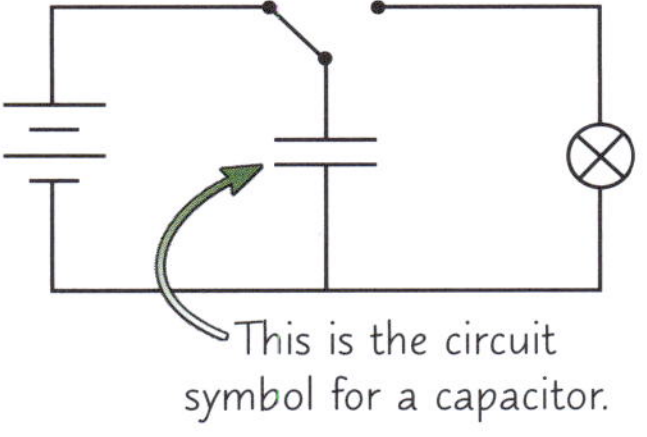

This is the circuit symbol for a capacitor.

1) In the circuit on the right, when the switch is flicked to the **left**, **charge** builds up on the plates of the **capacitor**. Energy provided by the battery is **stored** by the capacitor, in the form of **electrical potential energy**.
2) If the switch is flicked to the **right**, the energy stored on the plates will **discharge** through the **bulb**, converting electrical potential energy into light and heat.
3) Whilst the capacitor is charging, **work** is done **removing charge** from **one plate** and depositing **charge** onto the other one. The energy for this must come from the **energy** stored in the **battery**, and is given by **charge × average p.d.**
4) You can find the **energy stored** by the capacitor from the **area** under a **graph** of **p.d.** against **charge stored** on the capacitor. The p.d. across the capacitor is **proportional** to the charge stored on it, so the graph is a **straight line** through the origin. The **energy stored** is given by the **yellow triangle**.
5) **Area of triangle = ½ × base × height**, so the energy stored by the capacitor is: $E = \frac{1}{2}QV$
6) You can rearrange $C = \frac{Q}{V}$ to get $Q = CV$.
   If you substitute this into the equation for the area under the graph you get **another equation** for the energy stored on a capacitor:

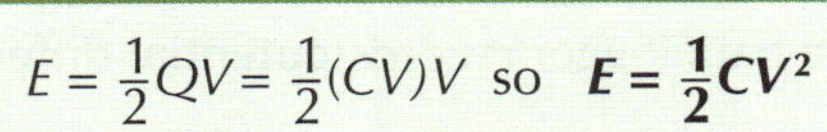

$$E = \frac{1}{2}QV = \frac{1}{2}(CV)V \quad \text{so} \quad E = \frac{1}{2}CV^2$$

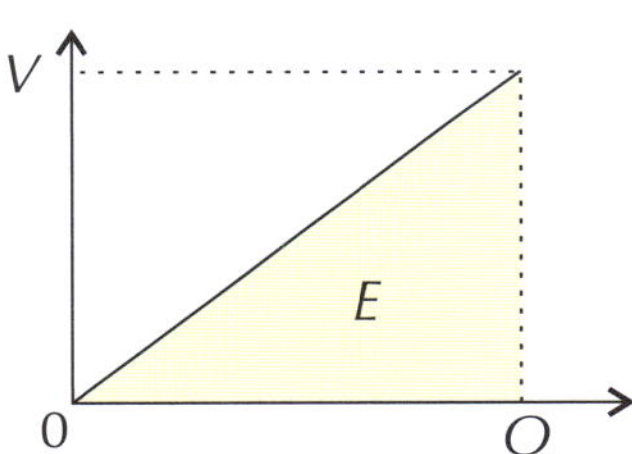

You may get given, or be asked to draw, a graph of charge stored against potential difference (i.e. a graph with the axes swapped). The energy stored is still the area under the graph, which will still be ½QV.

Practice Questions

Q1 Define capacitance.

Q2 How would you find the energy stored on a capacitor from a graph of potential difference against charge stored?

Exam Question

Q1 A 0.50 F capacitor is fully charged up from a 12 V supply.

a) What is the total energy stored by the capacitor:
A: 36 J B: 36 kJ C: 74 J D: 74 kJ [1 mark]

b) Calculate the charge stored by the capacitor. [1 mark]

Capacitance — fun, it's not...

Capacitors are really useful in the real world. Pick an appliance, any appliance, and it'll probably have a capacitor or several. If I'm being honest though, the only saving grace of this page for me is that it's not especially hard...

Charging and Discharging

You know, I've got the weirdest feeling I've seen some of these graphs before...

You can **Investigate** What Happens When you **Charge** a **Capacitor**

1) Set up the test circuit shown in the circuit diagram on the right.
2) Close the switch to connect the **uncharged** capacitor to the power supply.
3) Let the capacitor **charge** whilst the **data logger** records both the **potential difference** (from the voltmeter) and the **current** (from the ammeter) over time.
4) When the current through the ammeter is **zero**, the capacitor is fully charged (this won't take very long).
5) You should be able to plot the following graphs from the data collected from this experiment:

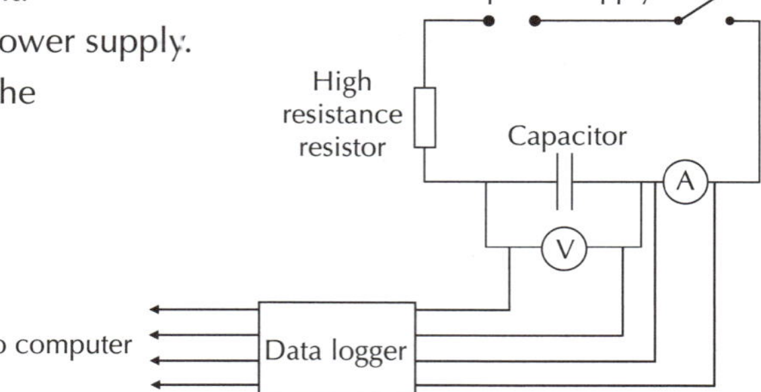

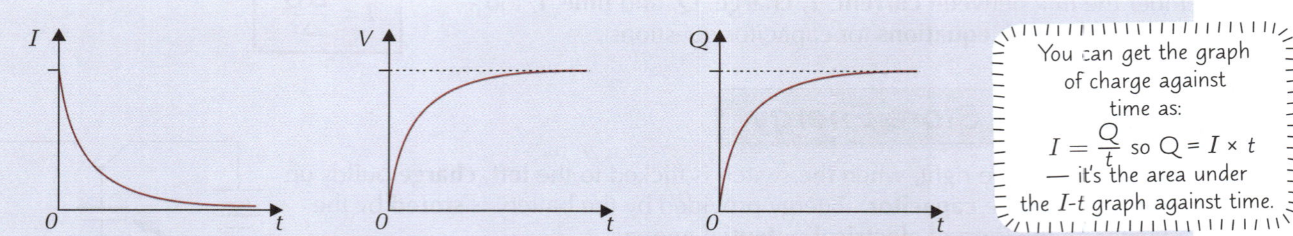

You can get the graph of charge against time as: $I = \frac{Q}{t}$ so $Q = I \times t$ — it's the area under the I-t graph against time.

1) As soon as the switch closes, current starts to flow. The electrons flow onto the plate connected to the **negative terminal** of the power supply, so a **negative charge** builds up.
2) This build-up of negative charge **repels** electrons off the plate connected to the **positive terminal** of the power supply, making that plate positive. These electrons are attracted to the positive terminal of the power supply.
3) An **equal** but **opposite** charge builds up on each plate, causing a **potential difference** between the plates. Remember that **no charge** can flow **between** the plates because they're **separated** by an **insulator** (e.g. a vacuum or air gap).
4) Initially, the current through the circuit is high. But, as **charge** builds up on the plates, **electrostatic repulsion** makes it **harder** and **harder** for more electrons to be deposited. When the p.d. across the **capacitor** is equal to the p.d. across the **power supply**, the **current** falls to **zero**.

To **Discharge** a Capacitor, **Disconnect** the **Power** and **Close** the **Switch**

1) **Disconnect** the power from the test circuit and close the **switch** to let the capacitor **discharge** whilst the data logger records **potential difference** and **current** over time.
2) When the **current** through the ammeter and the **potential difference** across the plates are **zero**, the capacitor is fully discharged.
3) You can then plot graphs of current, potential difference and charge against time once more.

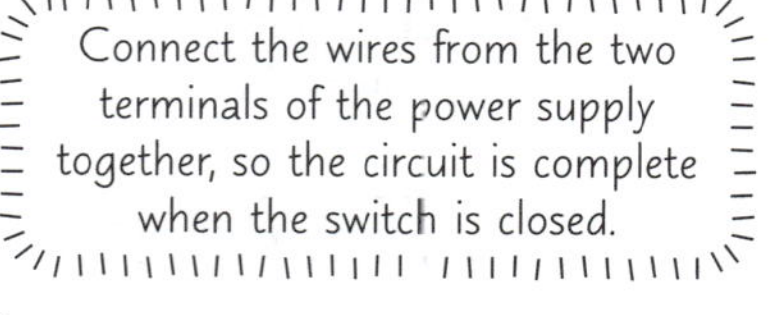

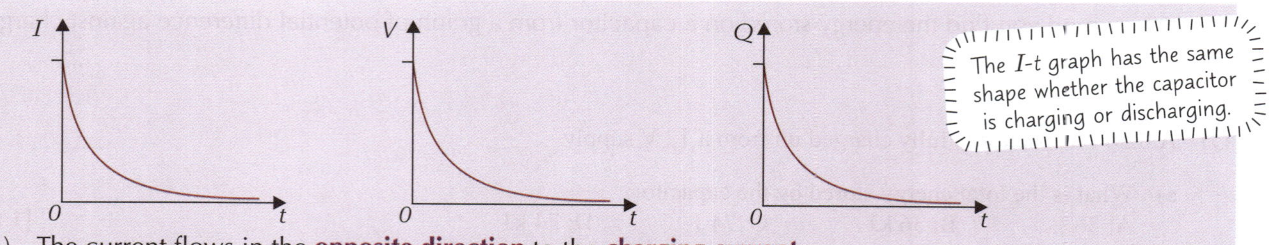

The I-t graph has the same shape whether the capacitor is charging or discharging.

1) The current flows in the **opposite direction** to the **charging current**.
2) As the **potential difference** decreases, the **current** decreases as well.
3) When a capacitor is **discharging**, the amount of **charge** on and **potential difference** between the plates falls **exponentially** with time. That means it always takes the **same length** of time for the charge or potential difference to **halve**, no matter what value it starts at — like radioactive decay (see p.5).
4) The same is true for the amount of **current flowing** around the circuit.

Charging and Discharging

The Time Taken to Charge or Discharge Depends on Two Factors

The **time** it takes to charge or discharge a capacitor depends on:

1) The **capacitance** of the capacitor (**C**). This affects the amount of **charge** that can be transferred for a given **potential difference**.
2) The **resistance** of the circuit (**R**). This affects the **current** in the circuit.

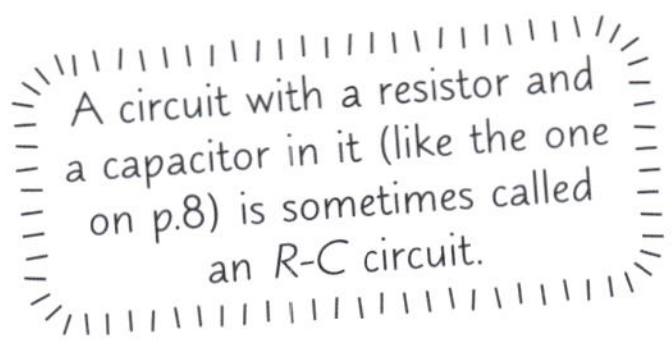

Discharge Rate is Proportional to the Charge Remaining

1) As you can see, the graph of **charge remaining** against **time** for a discharging capacitor has the same shape as the graph for radioactive decay (p.5). The amount of charge **initially falls quickly**, but the rate **slows** as the amount of **charge decreases** — the **rate of discharge** is **proportional** to the **charge remaining**.
2) You can show this **relationship** by drawing a **graph** of the rate of discharge (dQ/dt) against the charge remaining (Q) — you get a lovely straight line through the origin. This means that $\mathbf{dQ/dt \propto Q}$.
3) The rate of discharge of a capacitor also depends on the **capacitance** of the capacitor and the **resistance** of the circuit (see above). In fact, the rate of discharge of a capacitor is given by the **differential equation**:

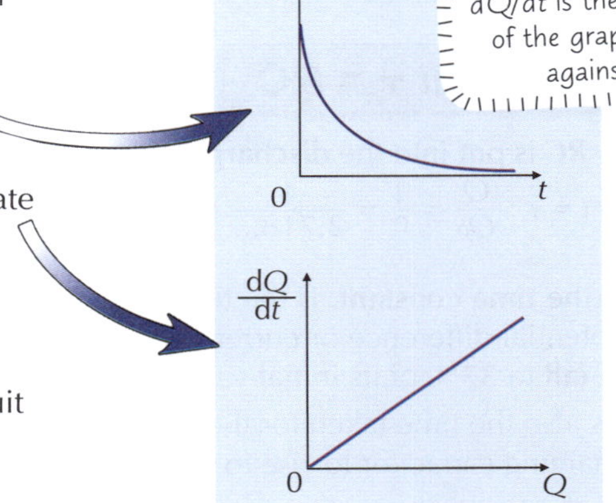

$$\frac{dQ}{dt} = -\frac{Q}{RC}$$

where Q is the charge remaining (C), R is the resistance of the circuit (Ω) and C is the capacitance of the capacitor (F)

4) The relationship between charge remaining and time for a discharging capacitor is an example of **exponential decay**. This means you can **model** the charge remaining on a discharging capacitor **in the same way** as you'd model radioactive decay.

You can Calculate Charge, P.d. and Current as a Capacitor Discharges

1) The **charge left** on the plates at a given time after a capacitor begins discharging from being fully charged is given by the equation:

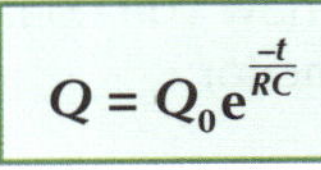

where Q_0 = the initial charge on the plates of the capacitor (C) and t = the time since the capacitor began discharging (s)

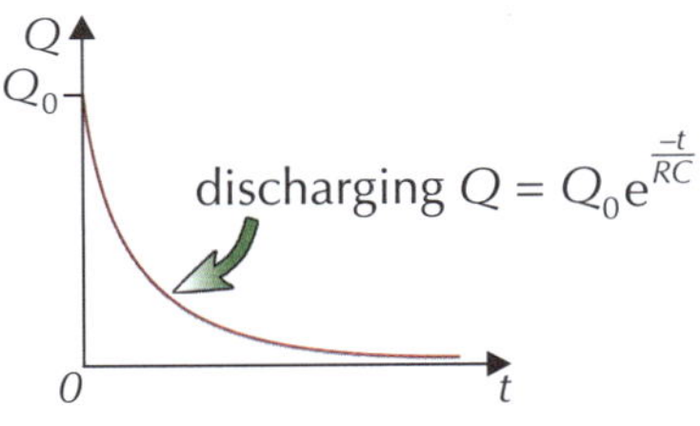

This is the **exact solution** of the differential equation $\frac{dQ}{dt} = -\frac{Q}{RC}$.

2) The **proportion** of charge that is transferred from a discharging capacitor in a given time interval is **constant** (so the same proportion of the remaining charge will be transferred in the first second as in the second second). This is called the **constant ratio property**, and is true for all exponential relationships.

In radioactive decay, the number of undecayed nuclei remaining halves in each half-life (p.6). This is also an example of the constant ratio property.

3) As the **potential difference** and **current** also decrease **exponentially** as a capacitor discharges, the formulas for calculating the current or potential difference at a certain time are similar:

$$I = I_0 e^{\frac{-t}{RC}} \qquad V = V_0 e^{\frac{-t}{RC}}$$

where I_0 = the initial current as the capacitor begins discharging, and V_0 = the initial potential difference across the capacitor

As these are also **exponential decay** relationships, the time taken for the **current** to **halve** and the time taken for the **potential difference** to **halve** are also **constant** for a **given capacitor** in a **given circuit**.

Charging and Discharging

You can do the Same for a *Charging Capacitor*

When a capacitor is **charging**, the **growth rate** of the amount of **charge on** and **potential difference across** the plates shows **exponential decay** (so over time they increase more and more slowly).

1) The **charge** on the plates at a given time after a capacitor begins charging is given by the equation: $\mathbf{Q = Q_0(1 - e^{\frac{-t}{RC}})}$

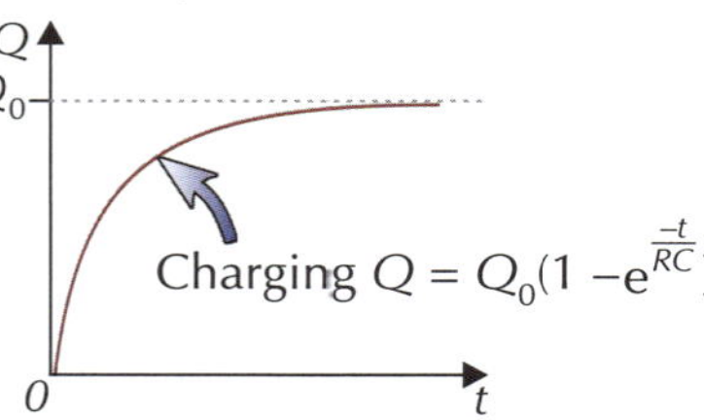

2) Similarly, the **potential difference** across the plates at a given time is given by the equation:

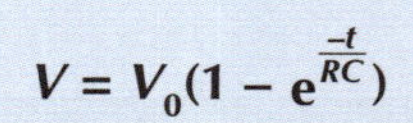

3) The **charging current** is different — it **decreases exponentially**. The formula to calculate the charging current at a given time is the same as for when it's discharging: $\mathbf{I = I_0 e^{\frac{-t}{RC}}}$

Time Constant τ = *RC*

τ is the Greek letter 'tau'

If $t = \tau = RC$ is put into the **discharging** equations on p.9, then $Q = Q_0e^{-1}$, $V = V_0e^{-1}$ and $I = I_0e^{-1}$.

So when $t = \tau$: $\frac{Q}{Q_0} = \frac{1}{e} = \frac{1}{2.718...} \approx 0.37$.

The units of τ are seconds, s.

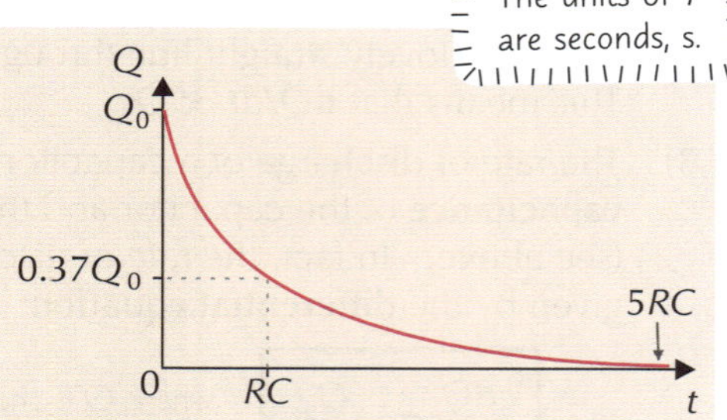

1) τ, the **time constant**, is the time taken for the charge, potential difference or current of a discharging capacitor to **fall** to **37%** of its initial value.
2) It's also the time taken for the charge or potential difference of a charging capacitor to **rise** to **63%** of its value when fully charged.
3) So the **larger** the **resistance** in series with the capacitor, the **longer** the capacitor takes to charge or discharge.
4) In practice, the time taken for a capacitor to charge or discharge **fully** is taken to be about 5*RC*.

Practice Questions

Q1 Describe an experiment to investigate how potential difference, current, and charge vary with time as a capacitor charges and discharges. Sketch graphs of the results you would expect to get.

Q2 What two factors affect how quickly a capacitor charges?

Q3 Write down the formula for calculating charge at a given time for a discharging capacitor.

Q4 Write down the formula for calculating potential difference at a given time for a charging capacitor.

Q5 What is the time constant of a discharging capacitor? How is it calculated? Describe how you could find the time constant of an *R-C* circuit from a graph of current against time for a charging capacitor.

Exam Questions

Q1 A 250 μF capacitor is fully charged from a 6.0 V battery and then discharged through a 1.0 kΩ resistor.

a) Calculate the potential difference across the capacitor 0.25 s after it begins charging. [2 marks]

b) Calculate the percentage of the total charge remaining on the capacitor 0.7 s after it begins discharging. [2 marks]

Q2 A fully charged 320 μF capacitor is discharged through a 1.6 kΩ resistor.

a) Calculate the time taken for the charge on the capacitor to fall to 37% of its fully charged value after it begins discharging. [2 marks]

b) Calculate the rate of change of the charge on the capacitor when the charge remaining is 5.5 mC. [1 mark]

I don't like capacitor circuits — they're so RC...

*You'll be given the equation for how the charge on a discharging capacitor changes with time in the exam, but you won't be given any of the other exponential equations on the last three pages (*gulp*), so make sure you learn them.*

Modelling Decay

Who'd have thought that capacitors and radioactive isotopes could have so much in common? Read on...

Capacitors and Radioactive Isotopes Have Similar Decay Equations

Radioactive isotopes (p.4-6) might seem very different from **capacitors** in *R-C* circuits (p.8-10), but the **decay models** for them are very similar. This **table** shows the **similarities** and **differences** between the models.

	DISCHARGING CAPACITORS	RADIOACTIVE ISOTOPES
1)	Decay equation is $Q = Q_0 e^{-t/RC}$.	Decay equation is $N = N_0 e^{-\lambda t}$.
2)	The **quantity** that decays is Q, the amount of charge left on the plates of the capacitor.	The **quantity** that decays is N, the number of undecayed nuclei remaining.
3)	**Initially**, the charge on the plates is Q_0.	**Initially**, the number of undecayed nuclei is N_0.
4)	It takes *RC* seconds for the amount of charge remaining to fall to **37% of its initial value**.	The time taken for the number of undecayed nuclei to decay by half (the **half-life**) is $T_{1/2} = \ln 2 / \lambda$.

You Can Use a Logarithmic Graph to Find the Decay Constant and Half-life

1) If you plot a **graph** of the **number of undecayed nuclei**, N, in a radioactive sample against **time**, t, you get an **exponential curve** like the one on page 5.
2) But, if you plot the **natural log** (ln) of the number of **undecayed nuclei** against **time**, you get a **straight line**. To find the **natural log** of a number, just use the **ln button** on your calculator.
3) You get a straight line because the **decay equation**, $N = N_0 e^{-\lambda t}$, can be **rearranged**, via the mystical wonder of **logs**, to the **general form** of a **straight line** — $y = mx + c$.
4) The **gradient** of the line is $-\lambda$ (the **decay constant**). From this you can **calculate** the **half-life**, $T_{1/2}$, of the sample.
5) This works for graphs of **activity** against **time** too — $A = A_0 e^{-\lambda t}$ becomes $\ln(A) = -\lambda t + \ln(A_0)$. But remember to **subtract** the **background activity** from any measurements first.

$$N = N_0 e^{-\lambda t} \Rightarrow \ln(N) = -\lambda t + \ln(N_0)$$
$$y = mx + c$$

If you're not sure about logs, take a look at the stuff on page 84 to see where this comes from.

Example: The graphs below show how the number of undecayed nuclei of a radioactive isotope decreases over time. Use each graph to calculate the half-life of the isotope.

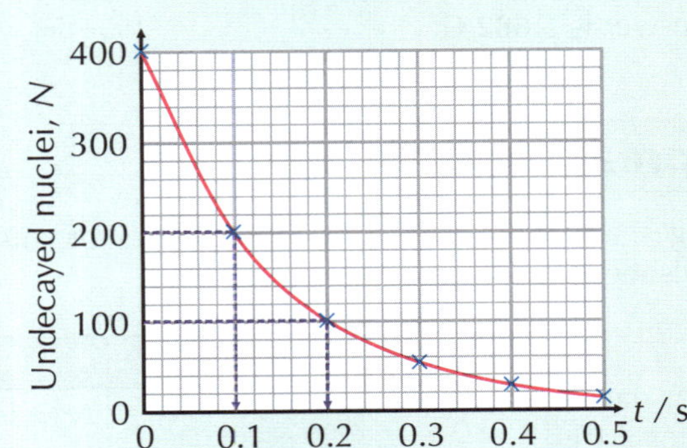

The first graph shows that it takes 0.1 s for the number of undecayed nuclei to fall from 400 to 200 (i.e. to halve).
So, the **half-life** is **0.1 s (to 1 s.f.)**.

The **gradient** of this graph is $-1.4 \div 0.2 = -7$, so the **decay constant**, $\lambda = 7$.
Substitute this into the equation for half-life:
$T_{1/2} = \ln(2) \div 7 = 0.099... =$ **0.1 s (to 1 s.f.)**.

You can use the **same method** with $Q = Q_0 e^{\frac{-t}{RC}}$ to find the **time constant** for an ***R-C* circuit** — the gradient of a graph of ln Q against t is $-\frac{1}{RC}$.

Modelling Decay

You Can Solve Differential Equations Using Iterative Methods

1) $\frac{dQ}{dt} = -\frac{Q}{RC}$ is a **differential equation** with an **exact solution**: $Q = Q_0 e^{\frac{-t}{RC}}$ (see p.9).
2) Many differential equations **don't have** exact solutions, so scientists use **iterative numerical methods** to solve them. Iterative methods work for **every type** of differential equation. However, the answers are only **approximate**.
3) Iterative methods work by breaking up the time period you're modelling over into **short intervals**, and calculating the value of the variable you're interested in at the end of each interval.
4) You can use the method below for **capacitor decay**, **radioactive decay** or **any** other model where the **rate of change** is **related** to the quantity that's changing — just substitute the **relevant equation**.

Before you start:

- Rearrange your equation to get dQ on its own: $dQ = -(Q/RC)dt$. You'll be dealing with **average changes** in Q rather than instantaneous ones, so you can rewrite this as $\Delta Q = -(Q/RC)\Delta t$ (see p.13).
- Pick a **time interval**, Δt, that the steps in your model will be separated by (the smaller the interval, the more accurate your answer will be, but the more iterations you'll need to do).

Then you're ready to go:

1) Start at time = 0, with the **initial value of Q**, (call this Q_0).
2) Increase the time by your time interval, Δt. **Substitute** your value of **Q** into the equation $\Delta Q = -(Q/RC)\Delta t$ to find ΔQ — the change in Q over this time interval.

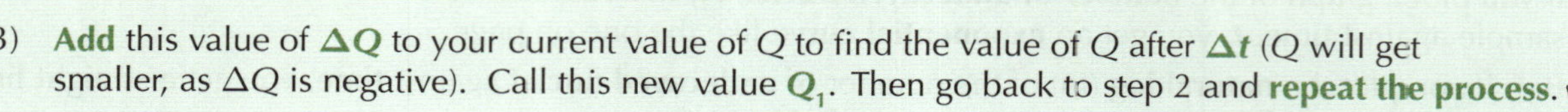

3) **Add** this value of ΔQ to your current value of Q to find the value of Q after Δt (Q will get smaller, as ΔQ is negative). Call this new value Q_1. Then go back to step 2 and **repeat the process**.

It's easiest to do this using a table — you'll need columns for time (*t*), charge (*Q*) and change in charge (ΔQ).

Example: A 150 μF capacitor storing 1.8 C of charge is discharged through a resistance of 40 kΩ. Using an iterative method and a time interval of 0.2 s, find the charge remaining on the capacitor after 0.4 s.

Just follow the method above:

1) Draw your table, and enter your **initial values** of Q (Q_0 = 1.8 C) and *t* (0 s).
2) **Add Δt** (0.2 s) to the initial time, and write this in the **second row** of the table.
3) Substitute the starting value of Q into the equation: $\Delta Q = -(Q/RC)\Delta t$ to find ΔQ over the first 0.2 seconds. Write this in the second row of the table.
4) **Add ΔQ** to the **initial value** of Q to find the value of Q at *t* = 0.2 s.
5) **Repeat the process** to complete the third row. You now have a value for Q after the time the question asks for (0.4 s) — so the answer is **1.682 C**.

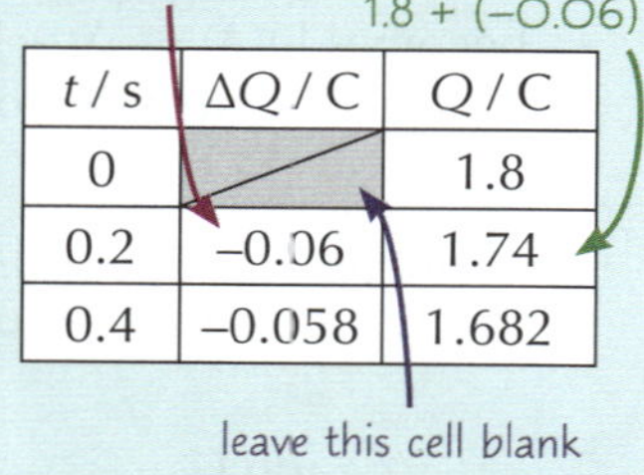

t / s	ΔQ / C	Q / C
0		1.8
0.2	−0.06	1.74
0.4	−0.058	1.682

You can Make Iterative Models using a Spreadsheet

Iterative methods involve lots of calculations, so they take ages, and there are lots of opportunities for you to make a **mistake**. You can get round this by using a **spreadsheet**.

To model how a sample of an isotope will decay:

1) Set up columns for **total time (*t*)**, **ΔN** and **N**, as well as data input cells for **Δt** and **λ**. Decide on a sensible value for **Δt**.
2) Rearrange $\frac{dN}{dt} = -\lambda N$ (p.4) to get $dN = -\lambda N dt$.
As you're calculating **average rates of change** in N (ΔN) over fixed time intervals, Δt, write this as: $\Delta N = -\lambda N \Delta t$.
3) **Enter formulas** into the spreadsheet to calculate the number of undecayed nuclei left in the sample after each time interval. If you write these properly, the spreadsheet can **automatically** fill them in for as many rows (iterations) as you want.

t / s	ΔN	N
$t_0 = 0$		N_0 = initial no. of nuclei
$t_1 = t_0 + \Delta t$	$(\Delta N)_1 = -\lambda N_0 \Delta t$	$N_1 = N_0 + (\Delta N)_1$
$t_2 = t_1 + \Delta t$	$(\Delta N)_2 = -\lambda N_1 \Delta t$	$N_2 = N_1 + (\Delta N)_2$
$t_3 = \dots$	$(\Delta N)_3 = \dots$	$N_3 = \dots$

λ (s⁻¹) = e.g. 1×10^{-4}
Δt (s) = e.g. 1000

Make sure the references to these cells don't change when you autofill new rows.

Modelling Decay

Iterative Methods Only Give You Approximate Answers

1) Whether you use a spreadsheet or do your calculations by hand, using an iterative method to model a relationship in the form **dx/dt = kx** only gives **approximate** answers.
2) This is because iterative methods assume that dx/dt **doesn't change** over the time interval you're using. In fact, dx/dt is changing **all the time** (you can tell this from the fact that relationships like this give you graphs that are smooth curves — remember dx/dt is the **gradient** of an *x*-*t* graph). dx/dt is called the **instantaneous rate of change**.
3) The **smaller** the value of Δt you use, the **more accurate** your model will be. But using smaller values of *t* means you need to do **more iterations**.
4) This is why making iterative models using **spreadsheets** that can carry out a very **large number** of iterations very **quickly** is really useful — it lets you make Δt as small as you need it to be.

Your model gives you an estimate of how much *x* changes by (Δx) over each time interval, Δt. You can use this to estimate the average rate of change of *x* over *t*, called $\Delta x/\Delta t$. As Δt gets smaller, $\Delta Q/\Delta t$ gets closer to dQ/dt.

'That's iteration sorted. Now, does this thing have solitaire?'

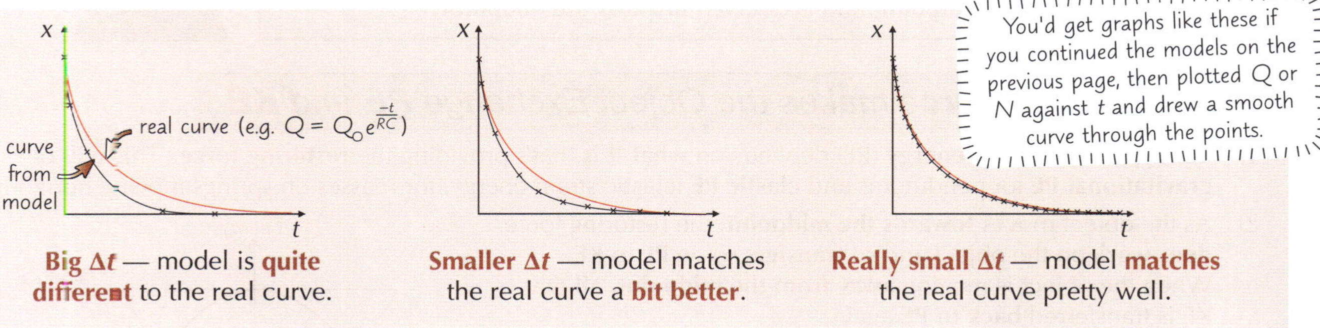

You'd get graphs like these if you continued the models on the previous page, then plotted Q or N against t and drew a smooth curve through the points.

Practice Question

Q1 Use a spreadsheet to model the radioactive decay of a sample initially containing 200 000 undecayed nuclei with a decay constant of 0.002 s^{-1}. If you don't have access to a computer, draw a table and do the first five iterations using a calculator.

Exam Questions

Q1 The graph on the right shows how the natural log of the activity of a sample of a radioactive isotope changes with time. The background activity has been subtracted from the graph.

a) Calculate the decay constant of the isotope. [1 mark]

b) Calculate the initial activity of the sample. [1 mark]

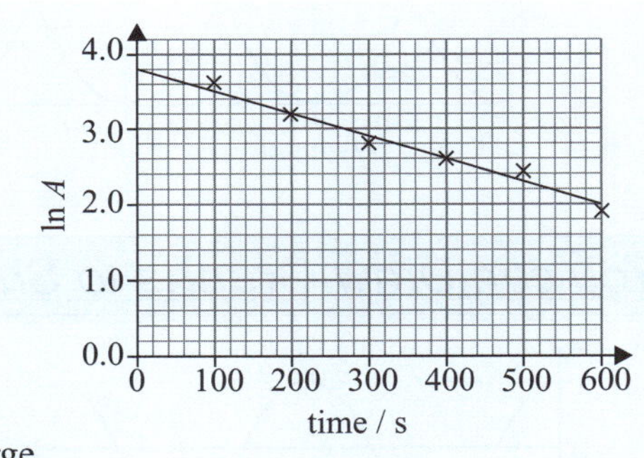

Q2 A teacher set up an *R*-*C* circuit to model the radioactive decay of an isotope of radon gas. The values of *R* and *C* were chosen so that the charge and resistance were related to the decay constant of the radon by $RC = \lambda^{-1}$. The capacitance of the capacitor was 520 μF and the resistance of the circuit was 144 kΩ. Calculate the decay constant of the radon. [2 marks]

Q3 A 10.0 μF capacitor is discharged through a resistance of 0.20 MΩ. The capacitor initially stores 5.0×10^{-2} C. Using an iterative method estimate the charge after 1.0 s. Use a time interval of 0.50 s for each iteration. [2 marks]

Modelling decay — when the train set in the back garden starts to rust...

Iteration is a bit of a pain, but it's really powerful — especially if you can get a computer to do all the tedious bits for you. This is the end of the capacitors and radioactive decay stuff (although radiation is coming up again later, you'll be pleased to know), so now's a good time to flick back over the last few pages to check it all makes sense...

Simple Harmonic Motion

Radioactive decay to capacitors to... simple harmonic motion? Well, of course — they're all examples of modelling.

SHM is Defined in Terms of Acceleration and Displacement

The **motion** of some **oscillating systems**, e.g. a **pendulum**, can be **modelled** by **simple harmonic motion** (SHM).

1) An object moving with **simple harmonic motion oscillates** to and fro, either side of a **midpoint**.
2) The distance of the object from the midpoint is called its **displacement**.
3) There is always a **restoring force** pulling or pushing the object back **towards** the **midpoint**.
4) The **size** of the **restoring force** depends on the **displacement**, and the force makes the object **accelerate** towards the midpoint:

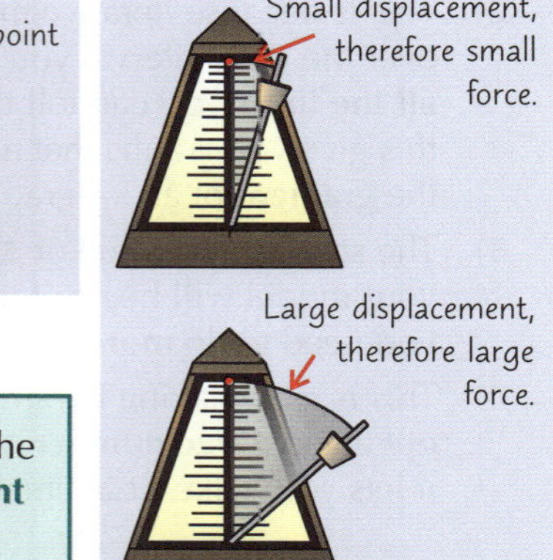

> **SHM:** an oscillation in which the **restoring force** on an object (and hence the **acceleration** of the object) is **directly proportional** to its **displacement** from the **midpoint**, and is directed **towards the midpoint**.

The Restoring Force makes the Object Exchange PE and KE

1) The **type** of **potential energy** (PE) depends on **what it is** that's providing the **restoring force**. This will be **gravitational PE** for pendulums and **elastic PE** (elastic strain energy) for masses on springs moving horizontally.
2) As the object moves **towards the midpoint**, the restoring force **does work** on the object and so **transfers** some **PE** to **KE**. When the object is moving **away from the midpoint**, all that KE is transferred **back to PE** again.
3) At the **midpoint**, the object's **PE** is **zero** and its **KE** is **maximum**.
4) At the **maximum displacement** (the **amplitude**) on both sides of the midpoint, the object's **KE** is **zero** and its **PE** is **maximum**.

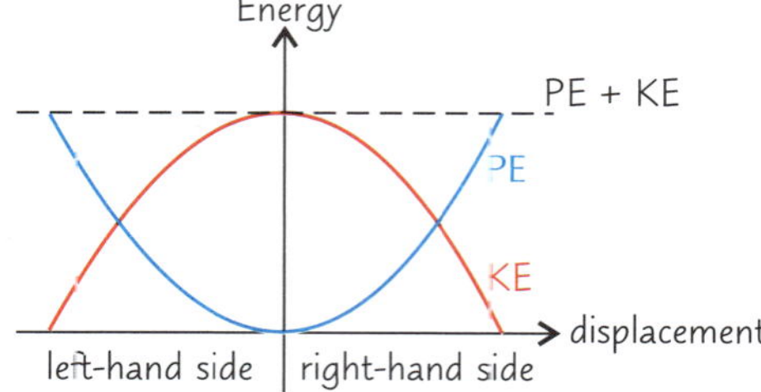

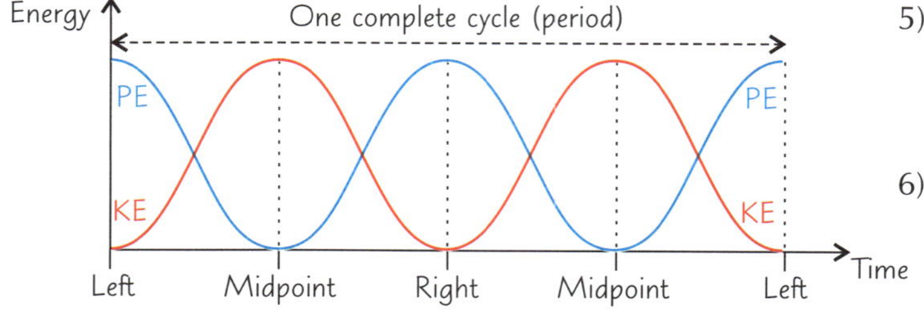

5) The **sum** of the **potential** and **kinetic** energy is called the **mechanical energy** and **stays constant** (as long as the motion isn't damped — see p.20-21).
6) The **energy transfer** for one complete cycle of oscillation (see graph) is: PE to KE to PE to KE to PE ... and then the process repeats... This is an example of the **conservation of energy**.

You can Draw Graphs to Show Displacement, Velocity and Acceleration

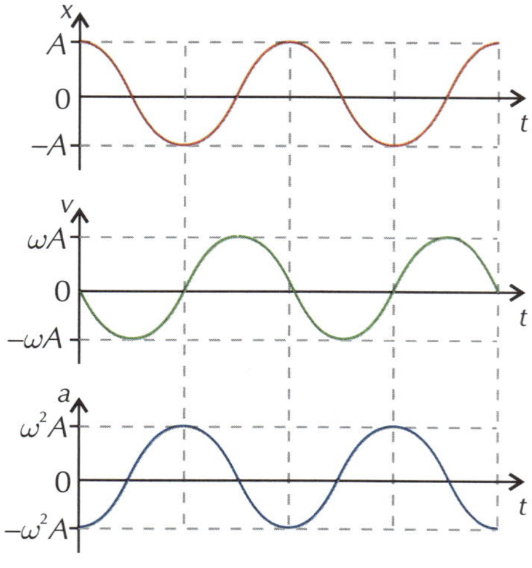

Displacement, x, varies with time, t, as a cosine (or sine) wave with a maximum value, A (the amplitude).

Velocity, v, is the gradient of the **displacement-time** graph. It is a **quarter of a cycle** in front of the **displacement** (a phase difference of $\pi/2$) and has a maximum value of ωA.

Acceleration, a, is the gradient of the **velocity-time** graph. It has a maximum value of $\omega^2 A$, and is in **antiphase** with the **displacement**.

ω is the **angular frequency** of the oscillation (in rad s^{-1}).

$$\omega = 2\pi f \quad \text{and} \quad \omega = \frac{2\pi}{T}$$

Where T is the time taken for one oscillation and f is the frequency (the number of oscillations per second — see the next page).

There's more on these graphs coming up on page 18.

Simple Harmonic Motion

The Frequency and Period don't depend on the Amplitude

1) From **maximum positive displacement** (e.g. maximum displacement to the right) to **maximum negative displacement** (e.g. maximum displacement to the left) and **back again** is called a **cycle** of oscillation.
2) The **frequency**, ***f***, of the SHM is the number of cycles per second (measured in Hz).
3) The **period**, ***T***, is the **time** taken for a complete cycle (in seconds).
4) The relationship between **frequency** and **period** is given by the equation: $f = \frac{1}{T}$
   In other words, they're **inversely proportional** to each other.

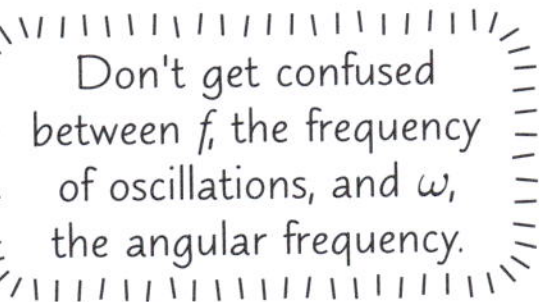

In SHM, the **frequency** and **period** are independent of the **amplitude** (i.e. constant for a given oscillation). So a pendulum clock will keep ticking in regular time intervals even if its swing becomes very small.

Learn the SHM Equations

1) According to the definition of SHM, the **acceleration**, $\mathbf{d^2x/dt^2}$, is directly proportional to the **displacement**, ***x***.
2) The **constant of proportionality** depends on the **frequency**, and the acceleration is always in the **opposite direction** to the displacement (so there's a minus sign in the equation).

$$\frac{d^2x}{dt^2} = a = -\omega^2 x$$

This is another differential equation. d^2x/dt^2 is the rate of change of dx/dt (ie the rate of change of the velocity — the acceleration).

3) The **velocity** is **positive** when the object's moving in one direction, and **negative** when it's moving in the opposite direction. For example, a **pendulum's velocity** is **positive** when it's moving from **left to right** and **negative** when it's moving from **right to left**.
4) The **displacement** varies with time according to one of two equations, depending on **where** the object was when the timing was started — they're really similar so you shouldn't have too much trouble learning both.

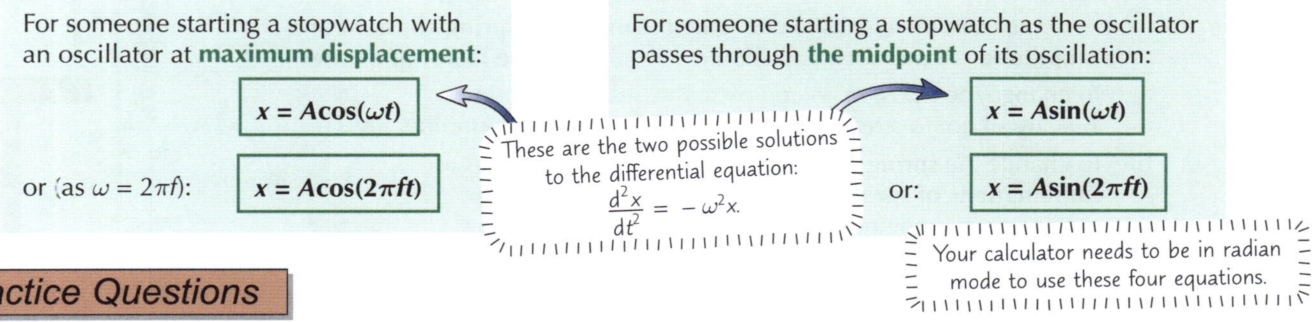

Your calculator needs to be in radian mode to use these four equations.

Practice Questions

Q1 What is the relationship between the acceleration and the displacement in SHM?

Q2 Sketch a graph of how the velocity of an object oscillating with SHM varies with time.

Q3 Describe how energy is transferred between kinetic and potential energy for a mass-spring system oscillating horizontally with SHM.

Q4 State the equation linking frequency and period.

Q5 Give two equations for the displacement of a simple harmonic oscillator, in terms of ω. When does each apply?

Exam Questions

Q1 a) Define simple harmonic motion. [2 marks]

b) Explain why the motion of a ball bouncing off the ground is not SHM. [1 mark]

Q2 A pendulum is pulled 0.050 m to the left and released. It oscillates with SHM with a frequency of 1.5 Hz.

a) What is the maximum velocity of the pendulum?
A: 0.24 ms^{-1} B: 9.4 ms^{-1} C: 0.15 ms^{-1} D: 0.47 ms^{-1} [1 mark]

b) Calculate its displacement 0.10 s after it is released. [2 marks]

"Simple" harmonic motion — hmmm, I'm not convinced...

The basic concept of SHM is simple enough (no pun intended). Make sure you can remember the shapes of all the graphs on page 14 and the equations from this page, then just get as much practice at using the equations as you can.

Investigating Simple Harmonic Motion

You can investigate simple harmonic motion in a few different ways — make sure you've read pages 14-15 before you tackle this lot, or it won't make much sense.

A Mass on a Spring is a Simple Harmonic Oscillator (SHO)

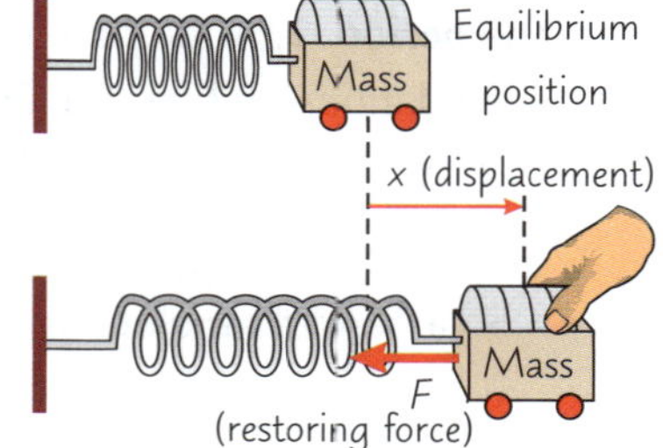

1) When the mass is **pushed to the left** or **pulled to the right** of the **equilibrium position**, there's a **force** exerted on it. The size of this force is:

$$F = kx$$

Where k is the spring constant (stiffness) of the spring in Nm^{-1} and x is the displacement in m. (This is Hooke's law, which you should have met in Year 1.)

2) After a bit of jiggery-pokery involving Newton's second law ($F = ma$) and some of the ideas on the previous page, you get the **formula for the period of a mass oscillating on a spring**:

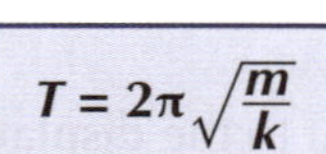

$$T = 2\pi\sqrt{\frac{m}{k}}$$

where T = period of oscillation in seconds, m = mass in kg
k = spring constant in Nm^{-1}

You can check this result **experimentally**:

1) Set up the equipment as shown in the diagram.
2) **Pull** the masses down a set amount — this displacement will be your initial **amplitude**. Let the masses go.
3) The masses will now oscillate with **simple harmonic motion**.
4) The **position sensor** will measure the **displacement** of the mass over **time**.
5) Connect the position sensor to a computer and create a **displacement-time** graph. Read off the period T from the graph.
6) You can measure the effects of changing the **mass**, the **spring constant** and the **amplitude**. Remember, you should only change **one factor at a time**.
   a) To change the **mass**, m — add extra masses to the spring (be careful not to stretch the spring past its limit of proportionality).
   b) To change the **spring stiffness**, k — use **different springs**, or **combinations of springs**. You can measure the stiffness of a spring by hanging a mass, m, from it and measuring how much it extends by, then using the equation $F = kx$, where $F = mg$ (i.e. the weight of the mass). If you use a **combination** of springs, you can **combine** their stiffnesses like this:
   c) To change the **amplitude**, A — pull the mass down by **different amounts** (again, be careful about how far you stretch the spring). Place a ruler behind the spring and line your eye up with the mass to make sure you measure the distance correctly (you could use a set-square to check that your eye is lined up with the ruler).
7) For each condition, you should take **repeated measurements**, and don't change any of the equipment halfway through the experiment.
8) You should get the following results:

a) $T \propto \sqrt{m}$ so $T^2 \propto m$

b) $T \propto \sqrt{\frac{1}{k}}$ so $T^2 \propto \frac{1}{k}$

c) T doesn't depend on amplitude, A.

Compressing or Stretching a Spring Stores Elastic Strain Energy

When the mass is **displaced** from the **equilibrium point**, the spring is **compressed** or **stretched**, so it stores **elastic strain energy**. You can work out the energy stored using:

$$E = \frac{1}{2}kx^2$$

You can also find the energy stored from a force-extension graph — the area under the graph is the elastic strain energy.

For a mass oscillating horizontally, this accounts for all of the potential energy in the system. For a mass oscillating vertically, you also need to take gravitational potential energy into account.

Investigating Simple Harmonic Motion

The Simple Pendulum is the Classic Example of an SHO

A **pendulum** will also move with simple harmonic motion, provided you don't displace it too far.
Here's a method for investigating the simple harmonic motion of a pendulum without using a computer:

1) Set up the equipment shown on the right (you'll also need a **stopwatch**). Measure the **mass**, ***m***, of the bob, and use a ruler to find the **length**, ***L***, of the string.
2) Pull the bob to the side (as shown) and measure the **angle**, ***A***, between the string and the vertical (it should be less than 10°).
3) Position your eye **level** with the **reference mark** (its called a fiduciary marker) then let the bob go. Start the stopwatch when the bob passes in front of the mark, then record the times when the bob passes in front of the mark again, travelling from the **same direction** (e.g. from left to right). The length of time that elapses between each time the bob passes in front of the mark from the same direction is the **time period**, ***T***, of the oscillator.
4) T might be **too short** to measure accurately from one swing. If so, measure the total time for a number of complete oscillations **combined** (say 5 or 10) and **divide** this time by the number of oscillations to find the time period.
5) You can then vary the **mass**, ***m***, of the bob, the **length**, ***L***, of the string, and the **angle**, ***A***, that you release the string from, keeping it less than 10° (changing this angle varies the amplitude of the oscillations). You should find that:
   a) $T \propto \sqrt{L}$, so $T^2 \propto L$ b) T does not depend on m c) T does not depend on A

Keeping your eye level with the mark will reduce errors in your results.

You could also do this experiment with an angle sensor and a data logger, which would give you more precise data.

In fact, for small angles of oscillation (up to about 10°), the **period of a pendulum** is:

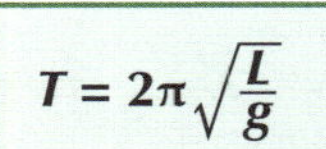

$$T = 2\pi\sqrt{\frac{L}{g}}$$

where T is the period of oscillation in s,
L is the length of the pendulum in m, and
g is the gravitational field strength in Nkg^{-1}

Practice Questions

Q1 Write down the formulas for the period of a mass on a spring and the period of a pendulum.

Q2 Describe a method you could use to measure the period of: a) a mass-spring system, b) a pendulum.

Q3 For a mass-spring system, what graphs could you plot to find out how the period depends on:
a) the mass, b) the spring constant, c) the amplitude?
What would they look like?

Exam Questions

Q1 A spring of original length 0.100 m is suspended from a stand and clamp.
A mass of 0.10 kg is attached to the bottom and the spring extends to a total length of 0.200 m.

a) Calculate the spring constant of the spring in Nm^{-1}. ($g = 9.81\ \text{Nkg}^{-1}$) [2 marks]

b) The mass is pulled down a further 0.020 m and then released.
The spring oscillates with simple harmonic motion, with a period of 0.63 s.
   i) Calculate the elastic strain energy stored in the spring when the mass is at the lowest point of its oscillation. [1 mark]
   ii) Calculate the mass that would be needed to make the period of oscillation twice as long. [2 marks]

Q2 Two pendulums of different lengths were released from rest at the top of their swing.
It took exactly the same time for the shorter pendulum to make five complete oscillations as it took the longer pendulum to make three complete oscillations.
The shorter pendulum had a length of 0.20 m. Show that the length of the longer one was 0.56 m. [3 marks]

Go on — SHO the examiners what you're made of...

The most important things to remember on these pages are those two period equations. You'll be given them in your exam, but you need to know what they mean and be happy using them.

Modelling Simple Harmonic Motion

Just when you thought the modelling was all over...

You Can Model Simple Harmonic Motion Graphically

1) If you carry out an **experiment** to investigate the simple harmonic motion of a mass-spring system using a **position sensor** and a computer (see page 16), you'll get data on how displacement changes with time, which you can get the computer to plot as a **graph**.

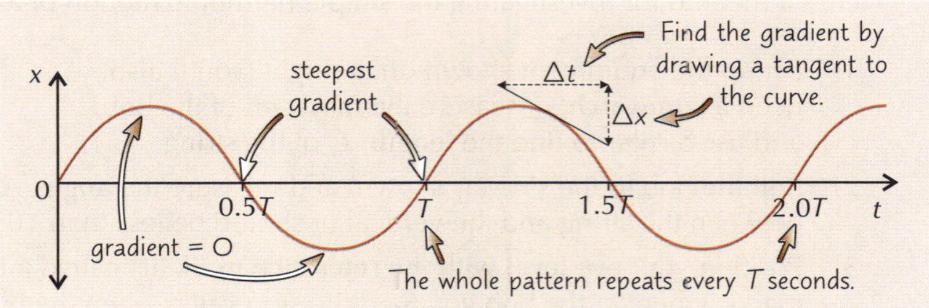

2) **Velocity** is the **rate of change** of **displacement** (dx/dt), so velocity is the **gradient** of an *x-t* graph. This means you can **generate** a graph of velocity against time for your data **without measuring velocity directly**. You'd need to find the gradient of the *x-t* graph at many points and plot these values against time. In practice you'd normally get the **computer** to calculate the velocity for you, but you can do this by **hand** if you need to.

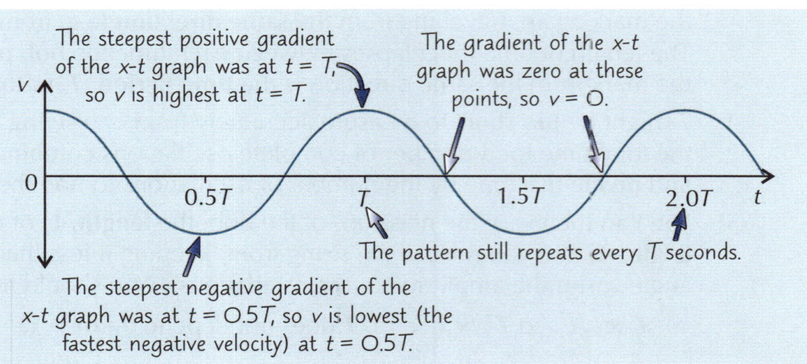

3) **Acceleration** is the **rate of change** of **velocity** (d^2x/dt^2). This means you can generate a graph of acceleration against time from a *v-t* graph by measuring the gradient of the *v-t* graph at many points and plotting these values against time.
The **shape** of the *a-t* graph is the **same** as the *x-t* graph, but ***a*** is **negative** when ***x*** is **positive** and vice versa. So $a \propto -x$ (which you know from page 15).

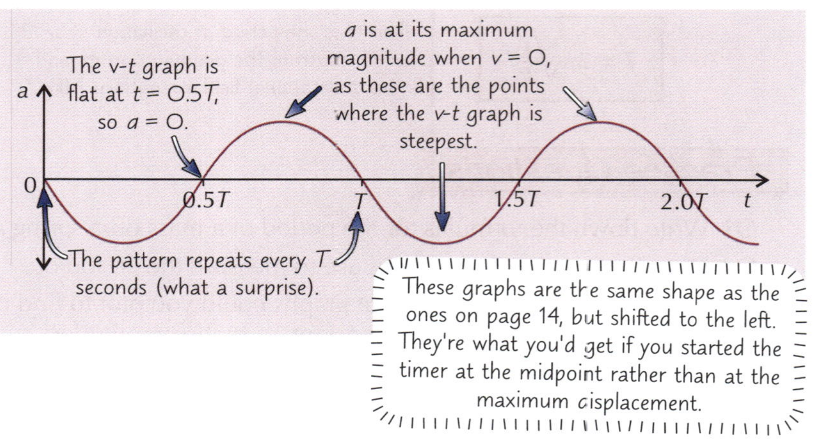

These graphs are the same shape as the ones on page 14, but shifted to the left. They're what you'd get if you started the timer at the midpoint rather than at the maximum displacement.

The Force on a SHO is Proportional to its Displacement

1) From the **definition** of **simple harmonic motion** (p.14) you know that the **restoring force** on a SHO is **proportional** to its **displacement**. You can express this in an equation as $F = -kx$. For a **mass on a spring**, *k* is the **spring constant**. For a **pendulum**, it's a bit more **complicated**. Just think of it as a constant.
2) Because $F = ma$, this means the **acceleration** of an SHO is **also proportional** to its **displacement**, as you can see in the graphs above.
3) Put all this together and you get:

$$F = ma = m\frac{d^2x}{dt^2} \quad \text{and} \quad F = -kx \quad \text{so:} \quad \boxed{\frac{d^2x}{dt^2} = -\frac{k}{m}x}$$

The minus sign in the equation $F = -kx$ means the force always acts to pull (or push) the oscillator back to the midpoint — it acts in the opposite direction to the oscillator's displacement.

4) This is a **differential equation** that you can use to model SHM. It's a bit like the ones for **radioactive decay** and **discharging capacitors** on p.4 and p.9, but it's $\frac{d^2x}{dt^2}$ rather than $\frac{dx}{dt}$. This is because it's an equation for the **rate of change of the rate of change** of *x* (i.e. the acceleration), rather than just the rate of change of *x*.

Modelling Simple Harmonic Motion

You Can *Model* SHM Using *Iteration*

Personally, I prefer modelling hats.

1) You can use the differential equation $\frac{d^2x}{dt^2} = -\frac{k}{m}x$ to **estimate** the **displacement** (or velocity) of an object moving with simple harmonic motion after a certain time.
2) You need to use an **iterative method** (like the one for capacitors and radioactive decay on page 12). As this equation is for the rate of change of the rate of change, your model will need a few **extra steps** — you'll need to model how the **velocity** of the oscillator changes over time to work out what happens to the **displacement**.

Example A 0.50 kg mass is attached to a spring with a spring constant of 10.0 Nm^{-1}. The mass is displaced by 0.050 m, then released and allowed to oscillate freely. Use an iterative method to find the displacement of the mass 0.2 s after its release. Use a time interval of 0.1 s for each iteration.

Before you start, you'll need to do some algebra:

- Remember, acceleration is the rate of change of velocity (v). So you can rewrite $\frac{d^2x}{dt^2} = -\frac{k}{m}x$ as $\frac{dv}{dt} = -\frac{k}{m}x$.
- As you're considering changes over a fixed time interval, change this to $\frac{\Delta v}{\Delta t} = -\frac{k}{m}x$ (p.13). Rearrange this to get Δv on its own: **Equation 1: $\Delta v = -(k/m)x\Delta t$**
- Velocity is change in displacement over time, $\frac{dx}{dt}$. So, over a fixed time interval, Δt, $v = \Delta x/\Delta t$. Rearrange this to get Δx on its own: **Equation 2: $\Delta x = v\Delta t$**

Now you can get going:

1) Draw a **table** like the one on the right and fill in the **time column** and the initial values of x and v. The initial velocity is **0 ms^{-1}** because the mass is held at a fixed displacement before being released.
2) Calculate Δv over the first time interval, using equation 1 above. Add this to the initial velocity to find v.
3) Calculate Δx using equation 2 above and this value of v. Add this to the initial displacement to find x after 0.1 s.
4) **Repeat the process** to complete the third row of the table. You now have a value for x after the time the question asks for (0.2 s) — so the answer is **0.022 m**.

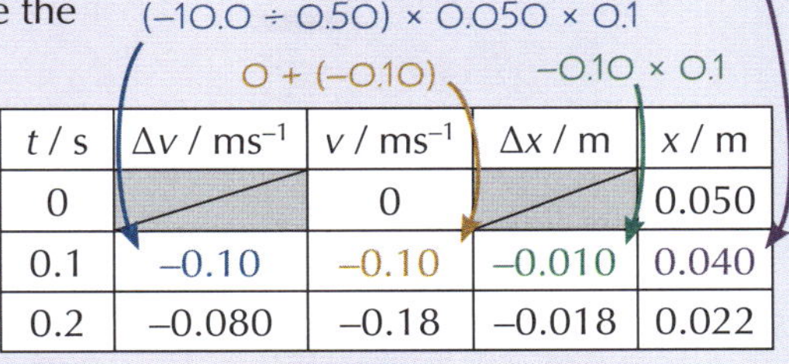

t / s	Δv / ms^{-1}	v / ms^{-1}	Δx / m	x / m
0		0		0.050
0.1	−0.10	−0.10	−0.010	0.040
0.2	−0.080	−0.18	−0.018	0.022

The acceleration at each iteration is just $\Delta v / \Delta t$.

As on page 12, you can make this kind of model using a spreadsheet. If you plotted the results of this model against time, you'd get sin and cos graphs like the ones on page 14. You could use these to read off the values of displacement or velocity at a given time.

Practice Questions

Q1 Explain how the graphs of displacement, velocity, and acceleration for simple harmonic motion are related to each other.

Q2 State the differential equation relating the acceleration of an SHO to its mass and displacement.

Exam Question

Q1 A 25 g pendulum bob is pulled 2.4 cm to the right and released. It swings with simple harmonic motion.

a) Using an iterative method, calculate the displacement of the bob after 0.15 s. Use a time interval of 0.05 s for each iteration (for this pendulum, $k = 0.82$ Nm^{-1}). [3 marks]

b) Calculate the maximum acceleration of the pendulum bob ($k = 0.82$ Nm^{-1}). [2 marks]

Iteration — it just seems to come up again and again...

The graph stuff here can be a bit painful, but getting your head round it will really help you to understand how displacement, velocity and acceleration are related to each other in simple harmonic motion. So if you're feeling a bit uncertain about what's going on, go back to the top of page 14 and have another look.

Free and Forced Vibrations

Resonance... tricky little beast. The Millennium Bridge was supposed to be a feat of British engineering, but it suffered from a severe case of the wobbles caused by resonance. How was it sorted out? By damping, of course — read on...

Free Vibrations — No Transfer of Energy to or from the Surroundings

1) If you stretch and release a mass on a spring, it oscillates at its **natural frequency**.
2) If **no energy's transferred** to or from the surroundings, it will **keep** oscillating with the **same amplitude forever**.
3) In practice this **never happens**, but a spring vibrating in air is called a **free vibration** anyway.
4) You need to know this formula for the **total energy** of a freely oscillating mass on a spring:
5) When x is at its maximum ($x = A$), all the energy is potential energy, so $E_{total} = \frac{1}{2}kA^2$.

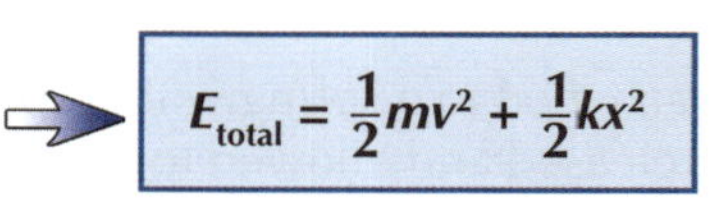

$$E_{total} = \frac{1}{2}mv^2 + \frac{1}{2}kx^2$$

or total energy = KE + PE

This is true for a mass oscillating horizontally. You'd need to include gravitational potential energy (p.25) if it was moving vertically (as in the example below).

Resonance Happens when Driving Frequency = Natural Frequency

1) A system can be **forced** to vibrate by a periodic **external force**.
2) The vibrations this produces are called **forced vibrations**.
3) The frequency of this force is called the **driving frequency**.
4) When the **driving frequency** approaches the **natural frequency**, the system gains more and more energy from the driving force and so vibrates with a **rapidly increasing amplitude**. When this happens the system is **resonating**.

Free and forced vibrations are also called free and forced oscillations.

Example: You can investigate how amplitude varies with driving frequency using a system like the one below.

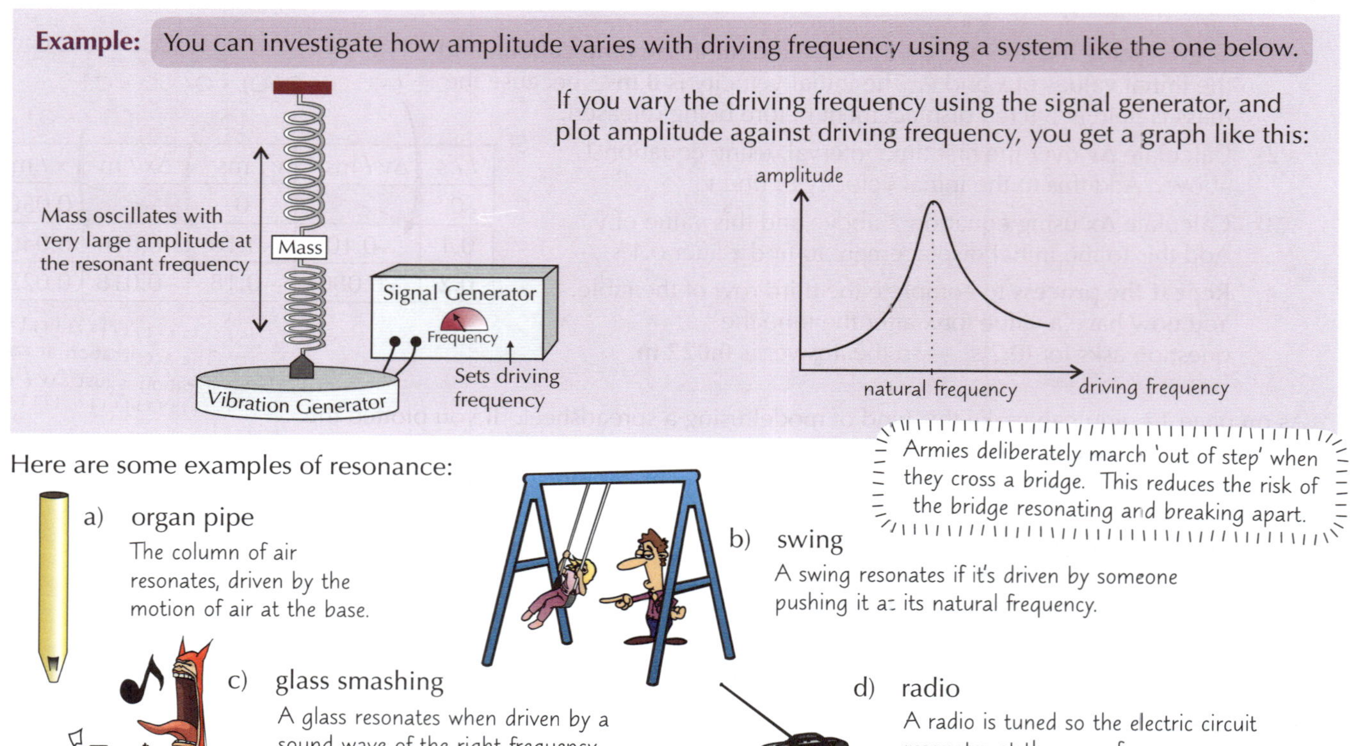

If you vary the driving frequency using the signal generator, and plot amplitude against driving frequency, you get a graph like this:

Armies deliberately march 'out of step' when they cross a bridge. This reduces the risk of the bridge resonating and breaking apart.

Here are some examples of resonance:

a) organ pipe
The column of air resonates, driven by the motion of air at the base.

b) swing
A swing resonates if it's driven by someone pushing it at its natural frequency.

c) glass smashing
A glass resonates when driven by a sound wave of the right frequency. This can make the glass break.

d) radio
A radio is tuned so the electric circuit resonates at the same frequency as the radio station you want to listen to.

Damping Happens when Energy is Lost to the Surroundings

1) In practice, **any** oscillating system **loses energy** to its surroundings.
2) This is usually down to **frictional forces** like air resistance.
3) These are called **damping forces**.
4) Systems are often **deliberately damped** to **stop** them oscillating or to **minimise** the effect of **resonance**.

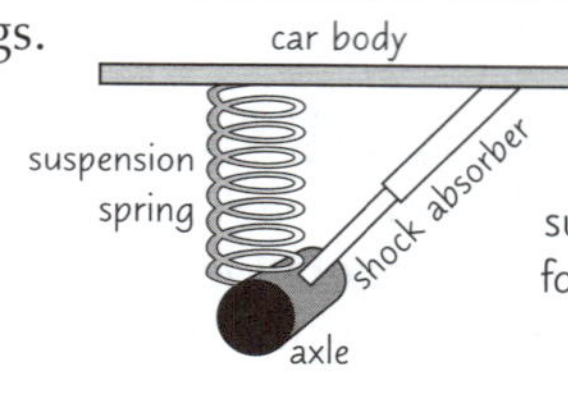

Shock absorbers in a car suspension provide a damping force by squashing oil through a hole when compressed.

Free and Forced Vibrations

Different Amounts of Damping have Different Effects

1) The **degree** of damping can vary from **light** damping (where the damping force is small) to **overdamping**.
2) Damping **reduces** the **amplitude** of the oscillation over time. The **heavier** the damping, the **quicker** the amplitude is reduced to zero.
3) **Critical damping** reduces the amplitude (i.e. stops the system oscillating) in the **shortest possible time**.
4) Car **suspension systems** and moving coil **meters** (which control the arm in analogue voltmeters and ammeters) are critically damped so that they **don't oscillate** but return to equilibrium as quickly as possible.
5) Systems with **even heavier damping** are **overdamped**. They take **longer** to return to equilibrium than a critically damped system.
6) **Plastic deformation** of ductile materials **reduces** the **amplitude** of oscillations in the same way as damping. As the material changes shape, it **absorbs energy**, so the oscillation will become smaller.

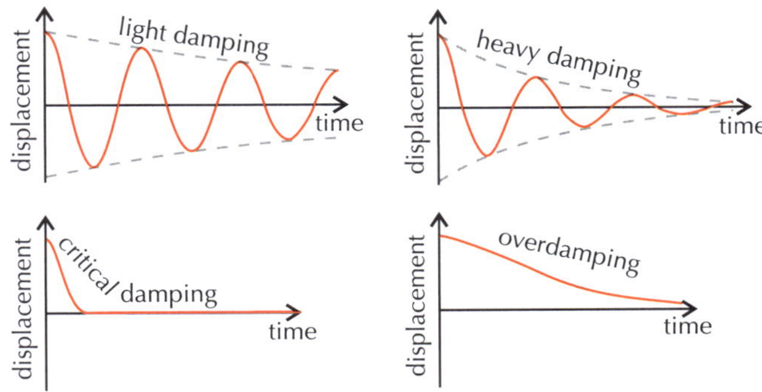

Damping Affects Resonance too

Structures are damped to avoid being damaged by resonance. Loudspeakers are also made to have as flat a response as possible so that they don't 'colour' the sound.

1) **Lightly damped** systems have a **very sharp** resonance peak. Their amplitude only increases dramatically when the **driving frequency** is **very close** to the **natural frequency**.
2) **Heavily damped** systems have a **flatter response**. Their amplitude doesn't increase very much near the natural frequency and they aren't as **sensitive** to the driving frequency.

Example: You can show how damping affects resonance using the experiment on the previous page.

Here's how increasing the damping affects resonance in this system:

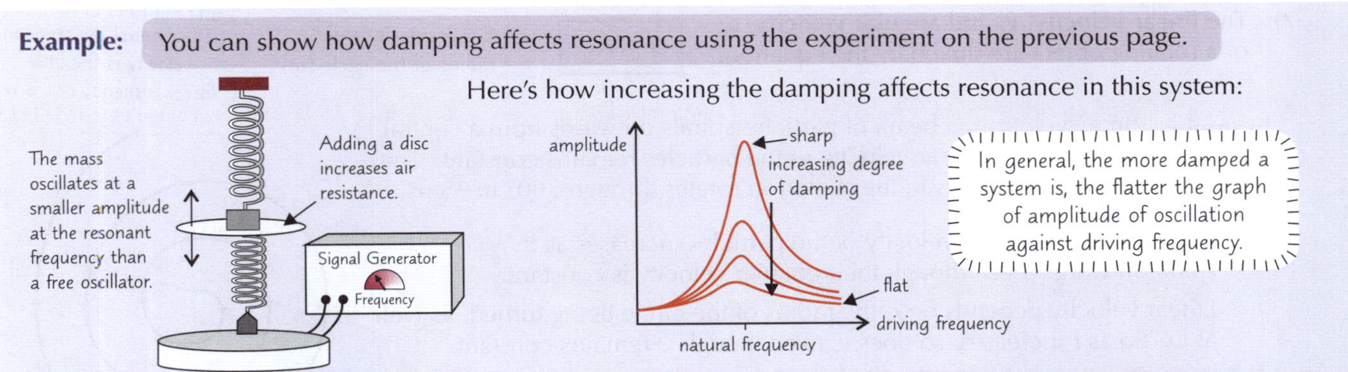

You could also investigate damping using a horizontal mass-spring system (top of p.16) by altering the surface (and hence the friction) below the trolley, or using a pendulum — damping will increase with the surface area of the bob.

Practice Questions

Q1 What is the equation for the total energy of a freely vibrating mass on a spring?

Q2 What is a free vibration? What is a forced vibration?

Q3 Define resonance and give an example of where it occurs.

Exam Questions

Q1 a) Draw a diagram to show how the amplitude of a lightly damped system varies with driving frequency. [2 marks]

b) On the same diagram, show how the amplitude of the system varies with driving frequency when it is heavily damped. [1 mark]

Q2 Define critical damping and state a situation where it is used. [2 marks]

Physics — it can really put a damper on your social life...

Resonance can be really useful (radios, organ pipes, swings — yay) or very, very bad...

Circular Motion

*It's probably worth putting a bookmark in here — this stuff is needed **all over** the place.*

Angles can be Expressed in Radians

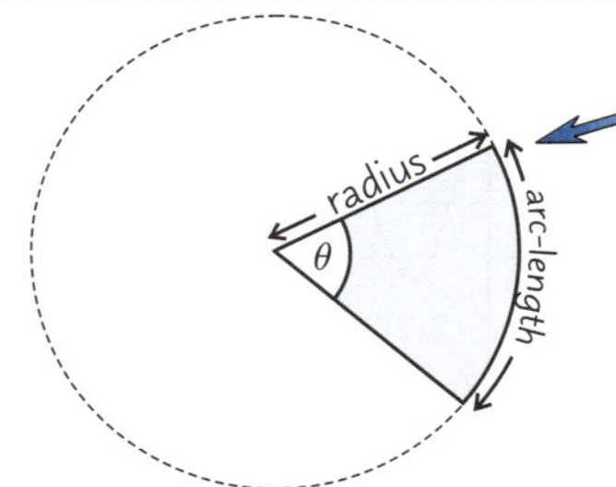

The angle in **radians**, θ, is defined as the **arc-length** divided by the **radius** of the circle, r.

For a **complete circle** (360°), the arc-length is just the circumference of the circle ($2\pi r$). Dividing this by the radius (r) gives 2π. So there are **2π radians in a complete circle.**

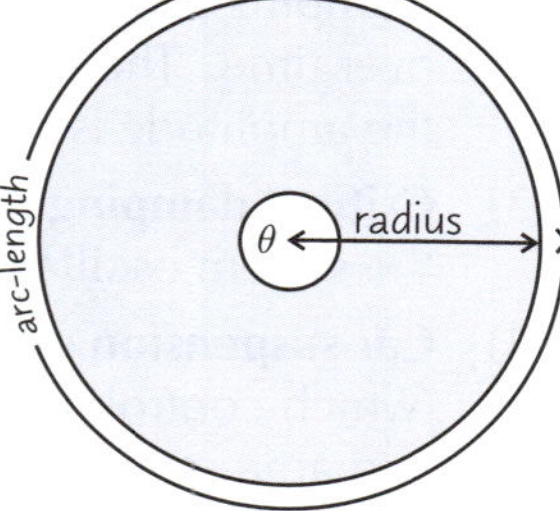

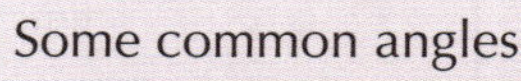

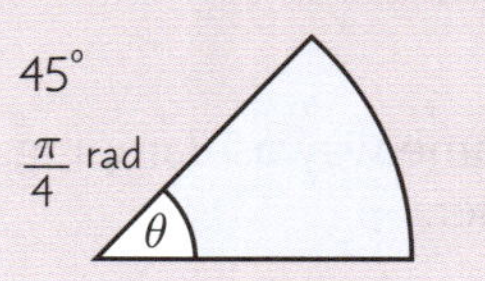

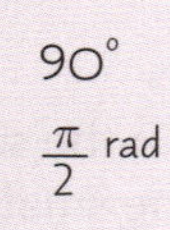

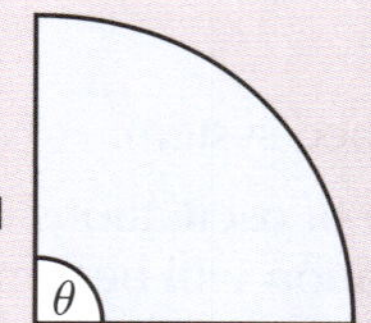

To convert from degrees to radians, multiply by $\frac{\pi}{180}$. To convert from radians to degrees, multiply by $\frac{180}{\pi}$. 1 radian ≈ 57°.

The Angular Velocity is the Angle an Object Rotates Through per Second

1) Just as **linear velocity**, v, is defined as displacement ÷ time, the **angular velocity**, ω, is defined as **angle ÷ time**. The unit is rad s^{-1} — radians per second.

$$\omega = \frac{\theta}{t}$$

ω = angular velocity (rad s^{-1})
θ = angle (rad) turned through in time t
t = time (s)

2) The **linear velocity**, v, and **angular velocity**, ω, of a rotating object are linked by the equation:

$$v = r\omega$$

v = linear velocity (ms^{-1})
r = radius of the circle (m)

The symbol for angular velocity is the little Greek 'omega', not a w.

Example: In a cyclotron, a beam of particles spirals outwards from a central point. The angular velocity of the particles remains constant. The beam of particles in the cyclotron rotates through 360° in 35 μs.

a) Explain why the linear velocity of the particles increases as they spiral outwards, even though their angular velocity is constant.

Linear velocity depends on r, the radius of the circle being turned, as well as ω. So as r increases, so does v, even though ω remains constant.

b) Calculate the linear velocity of a particle at a point 1.5 m from the centre of rotation.

First, calculate the angular velocity:

$$\omega = \frac{\theta}{t} = \frac{2\pi}{35 \times 10^{-6}} = 1.7951... \times 10^5 \text{ rad s}^{-1}$$

Then substitute ω into $v = r\omega$:

$$v = r\omega = 1.5 \times 1.7951... \times 10^5$$
$$= 2.6927... \times 10^5 \text{ ms}^{-1}$$
$$= \mathbf{2.7 \times 10^5 \text{ ms}^{-1} \text{ (to 2 s.f.)}}$$

You can find out more about cyclotrons on p.68.

Circular Motion has a Frequency and Period

1) The frequency, f, is the number of complete **revolutions per second** (rev s^{-1} or hertz, Hz).

2) The period, T, is the **time taken** for a complete revolution (in seconds). Frequency and period are **linked** by the equation:

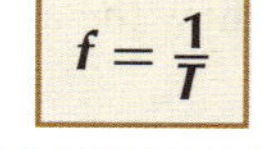

f = frequency in rev s^{-1} or Hz
T = period in s
ω = angular speed in rad s^{-1}

3) For a complete circle, an object turns through **2π radians** in a time T, so frequency and period are related to ω by:

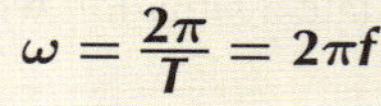

You should remember this second equation from p.14.

Circular Motion

An Object Travelling in a Circle is *Accelerating* since its *Velocity is Changing*

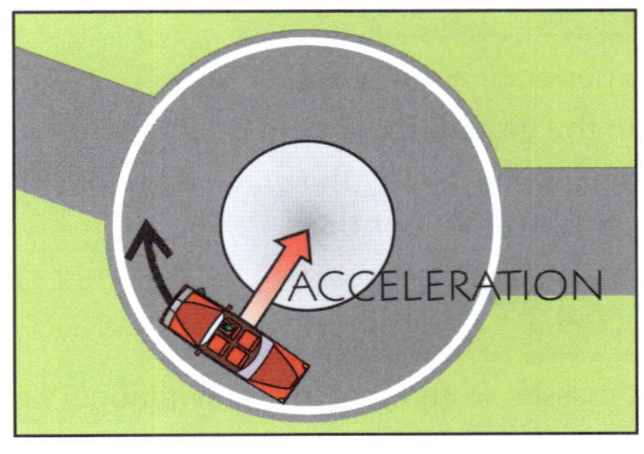

1) Even if the car shown is going at a **constant speed**, its **velocity** is changing since its **direction** is changing.
2) Since acceleration is defined as the **rate of change of velocity**, the car is accelerating even though it isn't going any faster.
3) This acceleration is called the **centripetal acceleration** and is always directed towards the **centre of the circle**.

There are two formulas for centripetal acceleration:

$$a = \frac{v^2}{r} \quad \text{and} \quad a = \omega^2 r$$

a = centripetal acceleration in ms^{-2},
v = linear velocity in ms^{-1},
ω = angular velocity in rad s^{-1}, r = radius in m.

The *Centripetal Acceleration* is produced by a *Centripetal Force*

From Newton's laws, if there's a **centripetal acceleration**, there must be a **centripetal force** acting towards the **centre of the circle**.

Since $F = ma$, the centripetal force must be:

$$F = \frac{mv^2}{r} \quad \text{and} \quad F = m\omega^2 r$$

F = centripetal force in N
m = mass in kg

The centripetal force is what keeps the object moving in a circle — remove the force and the object would fly off at a tangent.

Men cowered from the force of the centripede.

Practice Questions

Q1 How many radians are there in a complete circle?

Q2 How is angular velocity defined? What is the relationship between angular velocity and linear velocity?

Q3 Define the period and frequency of circular motion. What is the relationship between period and angular velocity?

Q4 Write equations for centripetal acceleration, a, and centripetal force, F, for an object of mass m, travelling at a linear velocity v, in a circular path with a radius r.

Q5 In which direction does the centripetal force act? What happens when this force is removed?

Exam Questions

Q1 The Earth orbits the Sun with an angular velocity of 2.0×10^{-7} rad s^{-1} at a radius of 1.5×10^{11} m. The Earth has a mass of 6.0×10^{24} kg.

a) Calculate the Earth's linear velocity. [2 marks]

b) i) Calculate the centripetal force needed to keep the Earth in its orbit. [2 marks]

ii) State what provides this centripetal force. [1 mark]

Q2 A car is driving in a circle around a roundabout with linear velocity 15 ms^{-1}. The radius of the car's circular motion is 12 m.

a) Calculate the centripetal acceleration of the car. [1 mark]

b) Calculate the time taken for the car to drive once around the roundabout. [2 marks]

My head is spinning after all that...

"Centripetal" just means "centre-seeking". The centripetal force is what actually causes circular motion. What you feel when you're spinning, though, is the reaction (centrifugal) force. Don't get the two mixed up.

Gravitational Fields

*Gravity's all about masses **attracting** each other. If the Earth didn't have a **gravitational field**, you'd be drifting around...*

Masses in a Gravitational Field Experience a Force of Attraction

1) A **gravitational field** is a force field — a **region** where an object will experience a **non-contact force**.
2) Any object with mass will **experience an attractive force** if you put it in the **gravitational field** of another object.
3) Only objects with a **large** mass, such as stars and planets, have a significant effect. E.g. the gravitational fields of the **Moon** and the **Sun** are noticeable here on Earth — they're the main cause of our **tides**.

You can Draw Field Lines to Show the Field Around an Object

Gravitational field lines are **arrows** showing the **direction of the force** that masses would feel in a gravitational field. The closer together the field lines are, the stronger the gravitational field.

1) A **uniform field** is a field that is **the same** everywhere. This is shown by the field lines being **equally spaced** and going in the **same direction**.
2) Point masses have a **radial field**. **Spherical** objects — like the **Earth** — can be **modelled** as point masses, as they act as if all of their mass is concentrated at their **centre**:

   1) The Earth's gravitational field is **radial** — the lines of force meet at its centre.
   2) If you put a small mass, ***m***, anywhere in the Earth's gravitational field, it will always be attracted **towards** the Earth. If you move mass ***m*** further away from the Earth — where the **lines** of force are **further apart** — the **force** it experiences **decreases**.
   3) The small mass, ***m***, has a gravitational field of its own. This doesn't have a noticeable effect on the Earth though, because the Earth is so much **more massive**.
   4) Close to the Earth's surface, the gravitational field is (almost) uniform — so the **field lines** are (almost) **parallel** and **equally spaced**. You can usually **assume** that the field is perfectly uniform.

You can Calculate Forces Using Newton's Law of Gravitation

The **force** experienced by an object in a gravitational field is always **attractive**. It's a **vector** which depends on the **masses** involved and the **distance** between them. It's easy to work this out for **point masses** using this equation:

Newton's Law of Gravitation:

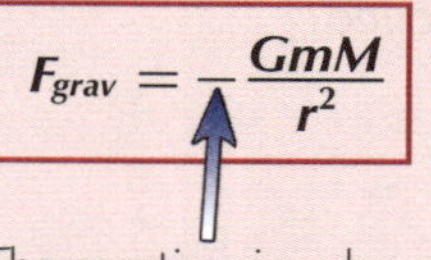

$$F_{grav} = -\frac{GmM}{r^2}$$

The negative sign shows that the vector F_{grav} is in the opposite direction to r (displacement of m from M).

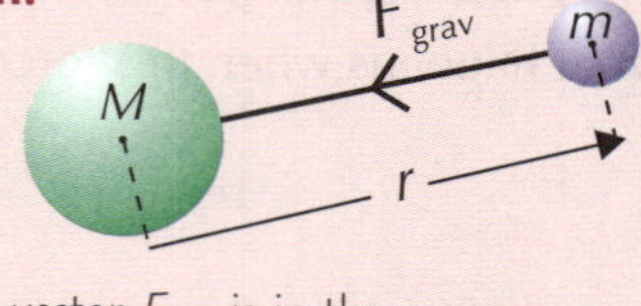

The diagram shows the force acting **on *m* due to *M***. (The force on M due to m is equal but in the opposite direction.) M and m behave as point masses. Both M and m are in kg. $G = 6.67 \times 10^{-11}$ Nm^{2}kg^{-2} is the **gravitational force constant**. r is the displacement (in metres) of m from M.

It doesn't matter what you call the masses: M and m, m_1 and m_2, Paul and Larry...

The law of gravitation is an **inverse square law** $\left(F_{grav} \propto \frac{1}{r^2}\right)$ so:

1) If the distance r between the masses **increases** then the force F_{grav} will **decrease**.
2) If the **distance doubles** then the **force** will be one **quarter** the strength of the original force.

The Field Strength is the Force per Unit Mass

Gravitational field strength, ***g***, is the **force per unit mass**. Its value depends on **where you are** in the field. There's a really simple equation for working it out:

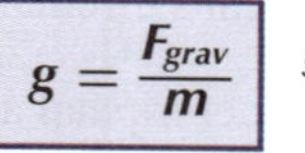

$$g = \frac{F_{grav}}{m}$$

g has units of newtons per kilogram (Nkg^{-1})

1) F_{grav} is the force experienced by a mass m when it's placed in the gravitational field at a given r. Divide F_{grav} by m and you get the **force per unit mass**.
2) g is a **vector** quantity, always pointing towards the centre of the mass whose field you're describing (because the gravitational force acts in that direction). This means g is often defined to be **negative**.
3) Since the gravitational field is almost uniform at the Earth's surface, you can assume g is a constant if you don't go too high.

   The **value** of g at the **Earth's surface** is approximately **9.81** Nkg^{-1} (or 9.81 ms^{-2}).

4) g is just the **acceleration** of a mass in a gravitational field. It's often called the **acceleration due to gravity**.

Gravitational Fields

A *Mass* in a *Uniform Gravitational Field* Experiences a *Constant Force*

The **force** on an object with mass m in a gravitational field is given by mg. In a **uniform gravitational field** (like near the Earth's surface) the value of g is the **same** at all locations. This means that the **force** experienced by a mass due to gravity will also be the **same** everywhere within the field.

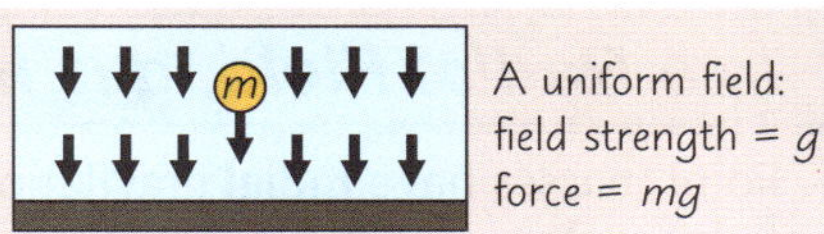

Potential Energy is *Proportional* to *Height* in a *Uniform Gravitational Field*

The change in **gravitational potential energy** of an object with mass m in a gravitational field is given by mgh, where h is the change in distance from a certain point — e.g. the **Earth's surface**.
In a uniform field, the value of mg is constant whatever the value of h, so:

1) **Increasing** the height of a mass, m, by a value of h changes the potential energy by the **same amount** wherever the mass is placed in the field. You can write this relationship as an **equation**:

$$\Delta E_{grav} = mgh$$

2) ΔE_{grav} is **positive** when the **height increases**, and **negative** when the **height decreases**.

You're only looking at the change in E_{grav} so you don't need to worry about the sign of g.

Energy is Never *Destroyed*

The **Principle of Conservation of Energy**: Energy **cannot be created** or **destroyed**. Energy **can be transferred** from one form to another but the total amount of energy in a closed system will not change.

Conservation of energy can be used to explain the changes in gravitational potential energy, mgh, and kinetic energy, $\frac{1}{2}mv^2$ when objects **move** in a **gravitational field**.

Example: A pomegranate of mass m is catapulted vertically with an initial velocity of 13 ms^{-1} from a height of 1.00 m above the Earth's surface. Calculate the height reached by the pomegranate, ignoring friction.

The kinetic energy given to the pomegranate is transferred into gravitational potential energy as the pomegranate gains height. The pomegranate will stop gaining height when all of the kinetic energy has been converted into gravitational potential energy: $\frac{1}{2}mv^2 = mgh$ so $h = \frac{v^2}{2g} = \frac{13^2}{2 \times 9.81} = 8.613...$ m

So the pomegranate reaches 8.613... + 1.00 = **9.6 m (to 2 s.f.)** above the Earth's surface.

Practice Questions

Q1 Describe and draw diagrams showing the Earth's gravitational field:
a) extending far from the Earth's surface, b) close to the Earth's surface.

Q2 Write down Newton's law of gravitation for two spherical masses, stating the assumption made about their fields.

Q3 What is meant by gravitational field strength?

Q4 What is the equation for a change in gravitational potential energy with a change in height h in a uniform field?

Exam Question

Q1 The Moon has a mass of 7.35×10^{22} kg and a radius of 1740 km.
($G = 6.67 \times 10^{-11}$ Nm^{2}kg^{-2})

a) Use Newton's law of gravitation to calculate the force on a 25 kg mass at the Moon's surface. [2 marks]

b) i) On the Moon, $g = -1.62$ Nkg^{-1}. A rock of 0.5 kg is fired vertically from the Moon's surface and reaches a maximum height of 4.5 m above the surface.
Assuming g is constant, calculate the initial velocity of the rock. [2 marks]

ii) Describe and explain the changes in gravitational potential and kinetic energy as the rock rises, changes direction and falls back to the surface. [3 marks]

If you're really stuck, put 'Inverse Square Law'...

Clever chap, Newton, but famously tetchy. He got into fights with other physicists, mainly over planetary motion and calculus... the usual playground squabbles. Then he spent the rest of his life trying to turn scrap metal into gold. Weird.

Gravitational Potential and Orbits

Gravity's a tricky little thing, don't you think? You know there must be a force pulling you to Earth (or whatever planet you're on), but you can't see it or feel it... or maybe you can, you just don't realise it because you've always felt it, hmm.

In a Radial Field, g *is* Inversely Proportional *to* r^2

Point masses have **radial** gravitational fields (see the diagram on page 24). The strength of the gravitational field, ***g***, **decreases** the **further away** you are from the centre of the mass, as shown on the graph below.

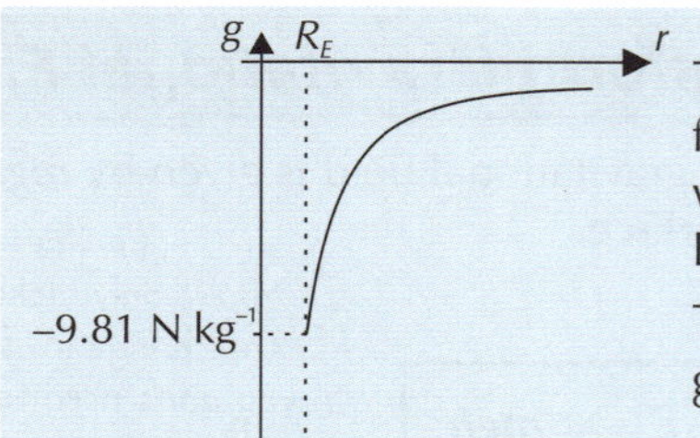

The graph shows how the gravitational field strength of the Earth, ***g***, varies with the distance from the centre of the Earth, ***r***. R_E is the Earth's radius.
The **area** under this curve gives you gravitational potential, ***V*** — see below.

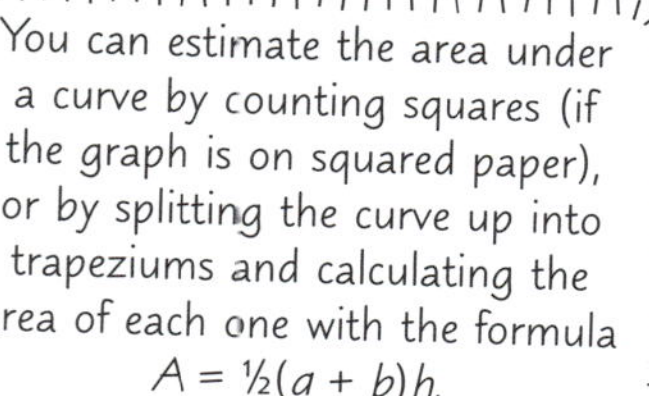
You can estimate the area under a curve by counting squares (if the graph is on squared paper), or by splitting the curve up into trapeziums and calculating the area of each one with the formula $A = \frac{1}{2}(a + b)h$.

You can see from the graph that the relationship between ***g*** and ***r*** isn't linear. In fact it's another **inverse square law** — make sure you know how to use it.

$$g = -\frac{GM}{r^2}$$

Remember, g has units N kg^{-1}.

You Gain Gravitational Potential Energy *if you Move Away from the Earth*

The **gravitational potential energy** of a mass at a certain point in a gravitational field, E_{grav}, is the **work** that would need to be done to move it from infinity to that point. Since gravitational fields are attractive, E_{grav} is always negative.

You can express the gravitational potential energy of a mass ***m*** in terms of its distance ***r*** from a large point mass ***M*** by combining the equation $E_{grav} = mgh$ (see previous page), where h is equal to r, with the one for ***g***, above.

$$E_{grav} = -\frac{GmM}{r}$$

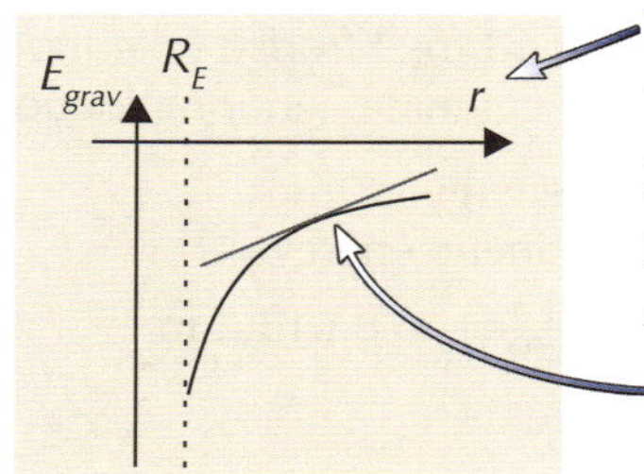

The graph shows how gravitational potential energy varies with distance from the Earth.

1) A mass on the Earth's surface (at R_E) has **negative** gravitational potential energy.
2) As you move a mass away from the Earth, it **gains potential energy**.
3) Potential energy is **zero** at an **infinite** distance from the Earth.
4) The gradient of a **tangent** to the graph gives the value of the gravitational **force** at that point — the force is **greatest** at the **Earth's surface** where the graph is **steepest**.

When you **move** a mass, m, between two different distances from the centre of a radial gravitational field, its gravitational potential energy will **change**. This is because **work is done** against the force of **gravity**. The work done depends on the **force** and the **distance moved**:

$$\Delta E_{grav} = F_{grav}\Delta s$$

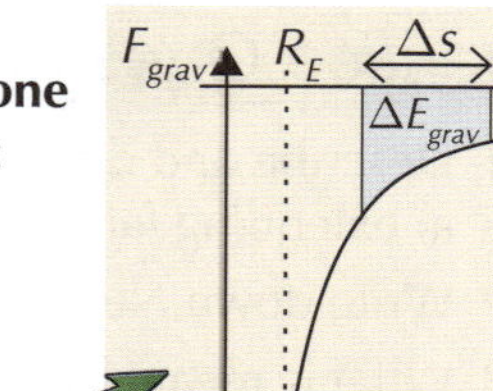

ΔE_{grav} can be found from a graph of F_{grav} **against** ***r***.

$\Delta E_{grav} = F_{grav}\Delta s$ = the **area** under the curve between the two distances.

If the mass m is moved but the distance from the mass M is kept **the same** (i.e. the mass is moved along an equipotential — see next page), no **work** is done against gravity ($\Delta s = 0$).

Gravitational Potential *is* Potential Energy per Unit Mass

The **gravitational potential** at a point, V_{grav}, is the **potential energy per unit mass**:

$$V_{grav} = \frac{E_{grav}}{m}$$

In a **radial field**, the equation is:

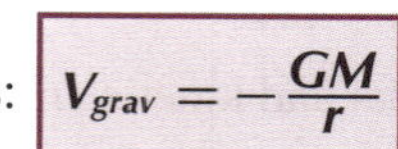

$$V_{grav} = -\frac{GM}{r}$$

The units of V_{grav} are either J kg^{-1} or N m kg^{-1}.

The graph of **gravitational potential** against distance has the **same shape** as the graph of **gravitational potential energy** against distance. Gravitational potential, V_{grav}, increases with distance, r, from the mass, and is **zero** at **infinity**.

The **gradient of a tangent** to the graph gives the value of ***g*** at that point.

Gravitational Potential and Orbits

Equipotentials Show All the Points in a Field Which Have the **Same** Potential

1) If you travel along a line of equipotential you **don't lose or gain energy** (**no** work is done).
2) For a uniform spherical mass (you can usually assume the Earth's one) the equipotentials are spherical surfaces.
3) **Equipotentials** and **field lines** are **perpendicular**.
4) At the Earth's surface, $V = -63\ \text{MJ kg}^{-1}$.

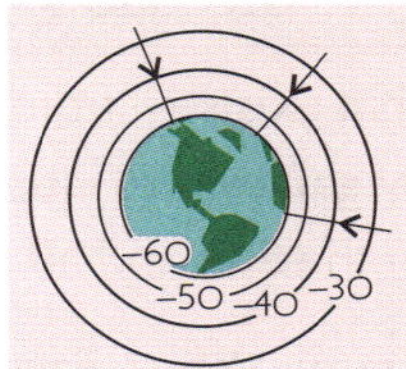

Equipotentials of –60, –50, –40 and –30 MJ kg^{-1} around Earth. Equipotentials at regular intervals aren't equally spaced — they get further and further apart (0 MJ kg^{-1} is at infinity)

Gravitational Fields Cause **Planetary Orbits**

1) A **satellite** is just any **smaller mass** that **orbits** a **much larger mass**.
2) For example, the **Moon** is a satellite of the Earth, the **planets** are satellites of the Sun and **man-made satellites** orbit the Earth (e.g. broadcasting TV signals).
3) In our Solar System, the planets have **nearly circular orbits**, so you can use the **equations of circular motion** to describe their motion — go back and have a look at pages 22-23 if you've forgotten them already.

1) Earth feels a force due to the gravitational 'pull' of the **Sun**. This force is given by Newton's law of gravitation:

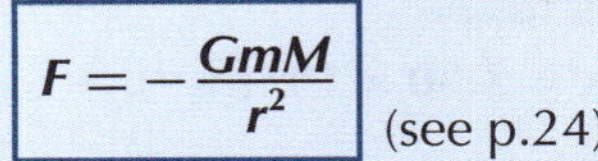

$$F = -\frac{GmM}{r^2}$$

(see p.24)

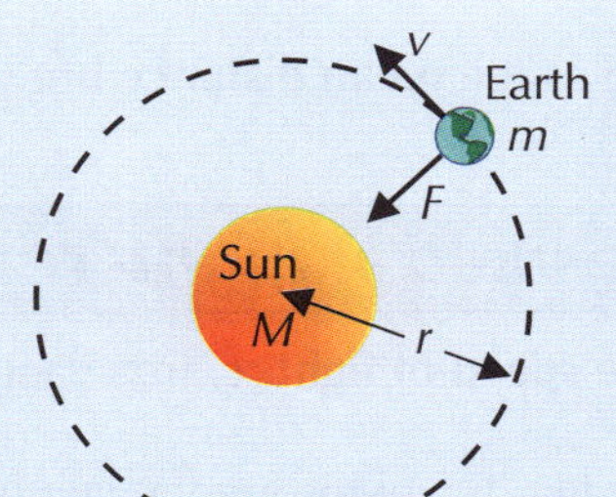

2) The Earth has velocity $\mathbf{v}$. Its linear speed is constant but its **direction** is not — so it's accelerating. The **centripetal force** causing this acceleration is:

$$F = \frac{mv^2}{r}$$

3) The **centripetal force** on the Earth must be a result of the **gravitational force** due to the Sun, and so these forces must be **equal**...

$$\frac{mv^2}{r} = \frac{GmM}{r^2} \quad \text{and rearranging...} \quad v = \sqrt{\frac{GM}{r}}$$

Practice Questions

Q1 Sketch graphs to show how the values of g and F_{grav} vary with distance from the Earth's surface.

Q2 What does the area under each graph you drew in Q1 represent?

Q3 What is the difference between gravitational potential and gravitational potential energy? Draw graphs of both quantities against distance for a radial field and state the quantity represented by the gradient of each graph.

Q4 Write down an equation for ΔE_{grav} in terms of F_{grav}.

Q5 Explain the meaning of equipotentials in terms of work done. What shape are the Earth's equipotentials?

Exam Question

(Use $G = 6.67 \times 10^{-11}\ \text{Nm}^2\text{kg}^{-2}$, mass of Earth = 5.98×10^{24} kg, radius of Earth = 6400 km)

Q1 A satellite with a mass of 3015 kg orbits 200 km above the Earth's surface.

a) Calculate the gravitational field strength at this distance from Earth. [2 marks]

b) Calculate the gravitational potential at this distance from Earth. [2 marks]

c) Calculate the satellite's gravitational potential energy. [1 mark]

d) Calculate the linear velocity of the satellite. [2 marks]

Increase your potential — stand on a chair...

It was a charming fellow called Johannes Kepler who showed that the planets' orbits are nearly circular. He's also been proclaimed as the first science fiction writer — busy chap. He wrote a tale about a fantastic trip to the Moon, where the book narrator's mum asks a demon the secret of space travel, to boldly go where — oh wait, different story.

Using Astronomical Scales

Space is big. I mean really, really big...

Our **Solar System** Contains the **Sun**, the **Planets** and Other Objects

1) The **universe** is **everything** that exists — from **stars** and **galaxies** to **microwave radiation** (page 33).
2) **Galaxies**, like our **Milky Way galaxy**, are clusters of **stars** and **planets** that are held together by gravity.
3) Our **Solar System** consists of the **Sun** and all of the objects that **orbit** it:

The planets (in order): **Mercury**, **Venus**, **Earth**, **Mars**, **Jupiter**, **Saturn**, **Uranus** and **Neptune** (as well as the asteroid belt between Mars and Jupiter) all have nearly **circular** orbits. **Pluto** is a dwarf planet beyond Neptune.

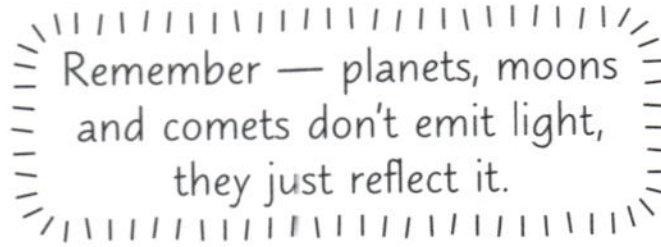

4) As well as planets, there are **moons**, **asteroids**, **comets**, **satellites** and lots of **dust** and **gas**.

Distances in the Solar System can be Measured in **Astronomical Units (AU)**

From the **16th century**, astronomers were able to work out the **distance** the **planets** are from the Sun **relative** to the Earth, using **astronomical units** (AU). But they could not work out the **actual distances** — it wasn't until 1769 that the size of the AU was accurately measured.

1 AU is about 150 million km.

One **astronomical unit** (AU) is defined as the **mean distance** between the **Earth** and the **Sun**.

Bigger Distances can be Measured in **Light-Years (ly)**

1) All **electromagnetic waves** travel at the **speed of light**, **c**, in a vacuum ($\mathbf{c = 3.00 \times 10^8}$ $\mathbf{ms^{-1}}$).

The **distance** that electromagnetic waves travel through a vacuum in **one year** is called a **light-year** (**ly**).

2) You need to know that 1 ly is around $\mathbf{10^{16}}$ **m**.

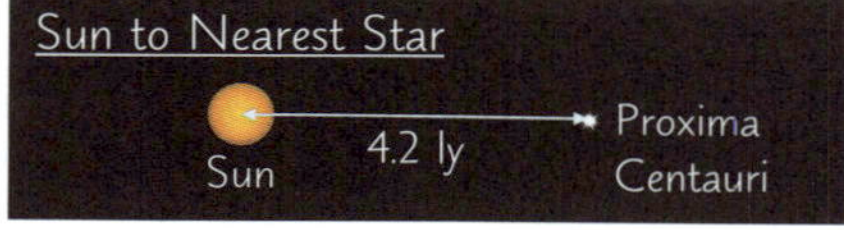

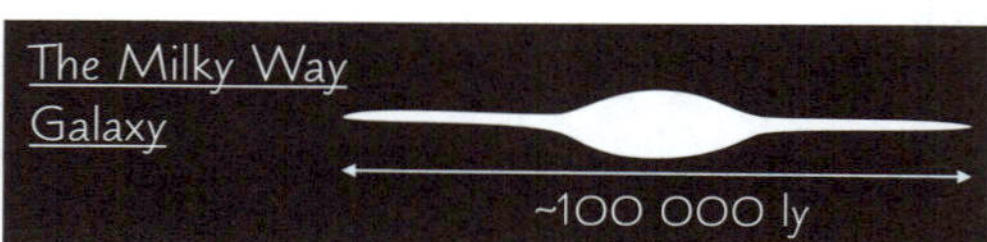

Just as a comparison, 1 ly is equivalent to about 63 000 AU.

3) If we see the light from a star that is, say, **10 light-years away** then we are actually seeing it as it was **10 years ago**. The further away an object is, the further **back in time** we are actually seeing it.
4) So when we look at the stars we're looking **back in time**, but since the universe has a **finite** age, we can only see as far back as the **beginning of the universe**. Since light has a finite **speed**, the **size** of the **observable universe** (the universe we can see) is limited (p.32).

Logarithmic Scales of Magnitude are used for **Large Ranges**

1) Quantities in astronomy can vary over huge ranges. For example, the distance to the Sun's **nearest star** is **hundreds of thousands** of times further than the Earth-Sun distance. And the distance to the furthest galaxy discovered so far is nearly **1 thousand trillion** times more than the Earth-Sun distance...
2) **Linear scales** (that go up in equal amounts) aren't much good for dealing with such huge ranges — **logarithmic scales** are often needed instead.
3) Logarithmic scales are **non-linear**. **Instead** of going up by equal amounts, each step on the scale is 'so many **times**' bigger than the last (e.g. **10 times**).

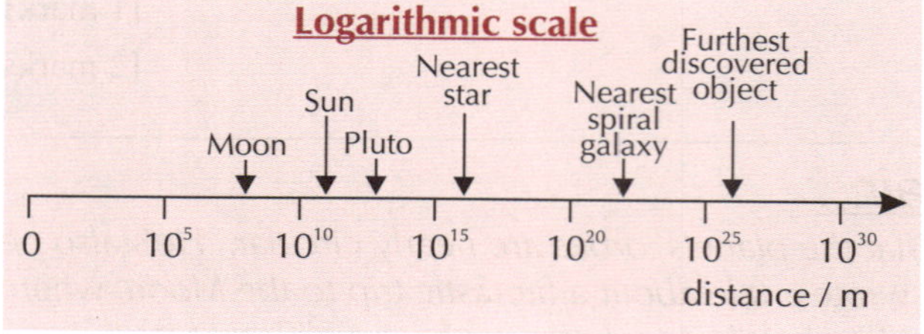

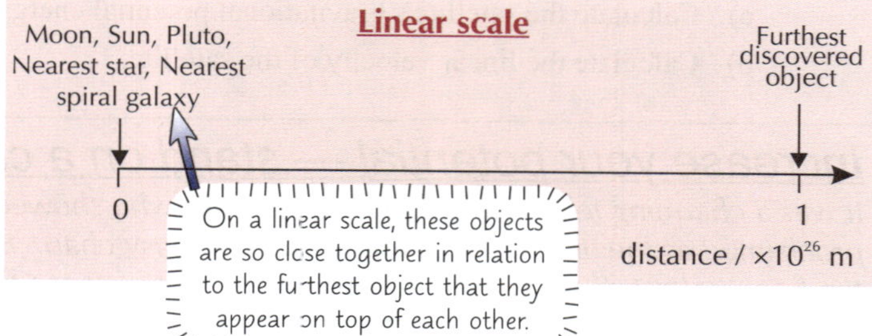

Using Astronomical Scales

There Are Loads of Examples of **Logarithmic Scales** in Astronomy

1) The **brightness** of stars is often measured with a logarithmic scale, called **apparent magnitude**.
2) Images of astronomical objects are often **displayed** using a logarithmic scale of **brightness**, so that the **difference** between the brightest and dimmest parts of the image is less intense.
3) The **Hertzsprung-Russell (H-R) diagram** is a plot of luminosity (power output) against temperature for stars. It has logarithmic scales on both axes.

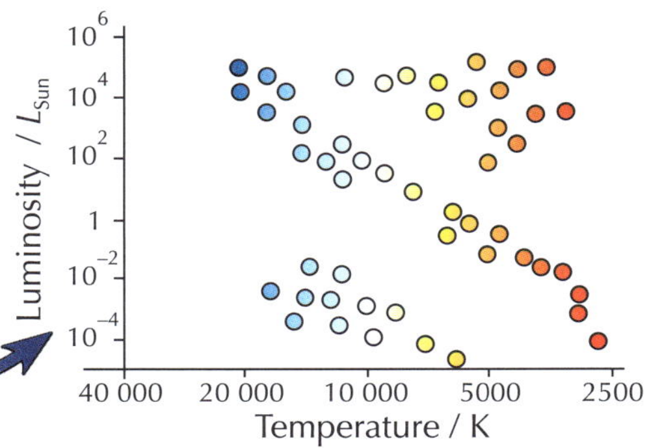

The **H-R Diagram** Can Be Used to Find the **Ages** of Stars

Finding the **age** of objects in astronomy is a tricky process. For an **individual star**, it is impossible to tell how **old** it is, even if you **know** properties like its distance and brightness.

Instead, astronomers need to study **clusters** — groups of stars that are assumed to have formed at the **same time**. They plot them on the **Hertzsprung-Russell (H-R) diagram** to work out what stages of their lives they're at:

1) Stars spend the **majority** of their lives fusing hydrogen into helium to produce energy (p.81). This is known as the **main sequence** stage. The **position** of a star on the H-R diagram main sequence depends on its **mass**.
2) At the **end** of this stage, the stars **move** off the main sequence to the **top-right** of the H-R diagram.
3) The **time spent** on the main sequence depends on the **mass** of the star — **more massive** stars use their fuel **more quickly** and spend **less time** in this stage. This means that the more **massive** stars at the top-left of the main sequence **leave** the main sequence **first**.
4) When astronomers plot the stars in a **cluster** on a **H-R diagram**, they can identify the **most massive** star left on the main sequence. They **assume** that it's about to **leave** the main sequence, and work out how old a star of that **mass** would be when it **uses up** all of its hydrogen fuel. This gives an estimate of the **age** of the star, and therefore the **age of the cluster**.

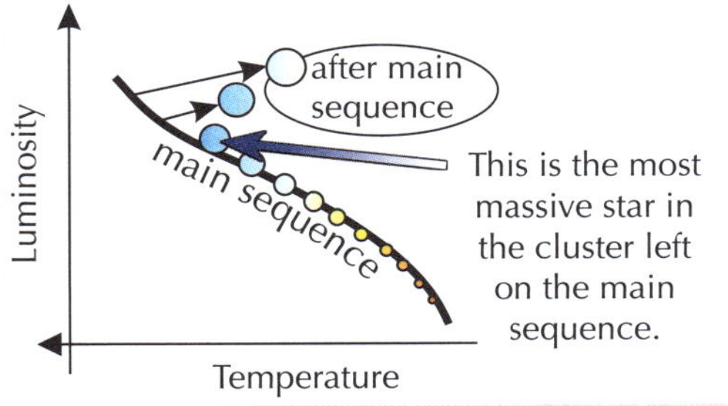

Practice Questions

Q1 What is the definition of an astronomical unit?

Q2 What is the difference between a linear scale and a logarithmic scale?

Q3 What assumption is made about a star cluster when determining the ages of the stars within it?

Exam Questions

Q1 The apparent magnitude, m, of a star is how bright it appears to be from Earth. It depends on the intensity of radiation received on Earth, I, and is defined using a logarithmic scale, where $m = -2.5 \log_{10} I + \text{constant}$.

a) Show that the intensity of a magnitude 1 star ($m = 1$) is $10^{\frac{2}{5}}$ greater than that of a magnitude 2 star. [3 marks]

b) State how many times greater is the intensity of a magnitude 1 star than that of a magnitude 6 star. [1 mark]

Q2 a) State the definition of a light-year. [1 mark]

b) Light travels at 3.00×10^8 ms^{-1} in a vacuum. Calculate the distance of a light-year in metres. [2 marks]

c) Explain why the size of the observable universe is limited by the speed of light. [2 marks]

So — using a ruler's out of the question then...

Good luck getting your head round these distances — they're pretty tricky to imagine. That's why astronomers often need to use logarithmic scales — imagine the Hertzsprung-Russell diagram on a linear scale and it should become clear.

Astronomical Distances and Velocities

There are loads of ways of measuring distances in space. Which one you use depends on how far away the object is. Here are a couple of ways to do it...

Distances *and* ***Velocities*** *in the Solar System can be Measured using* ***Radar***

1) **Distances** between objects in the **Solar System** are enormous — one way to measure them is using **radar**.
2) A **short pulse** of **radio waves** is sent from a **radio telescope** towards a distant object, e.g. a planet or asteroid. When the pulse hits the surface of the object, it's **reflected** back to Earth.
3) The telescope picks up the reflected radio waves and records the **time**, ***t***, taken for them to return.
4) Radio waves in space, like all electromagnetic waves, travel at the **speed of light**, ***c***, so you can work out the **distance**, ***d***, to the object using a variation of the formula **speed = distance ÷ time**:

$$2d = ct$$

It's $2d$, not just d, because the pulse travels twice the distance to the object — there and back again.

5) **Space-time worldlines** can also be used to find the distance to an object from the time interval between **sending** and **receiving** a radar pulse. On a grid of time (in seconds) against distance (in **light-seconds**), **start** the line showing the radar pulse at $t = 0$. The **gradient** of the line should be **1** (as light travels **1 light-second** in **1 second**). The radar pulse **changes direction** half way between leaving and arriving back, so the line should too. The point at which it changes direction is the **distance** of the object that reflected it.

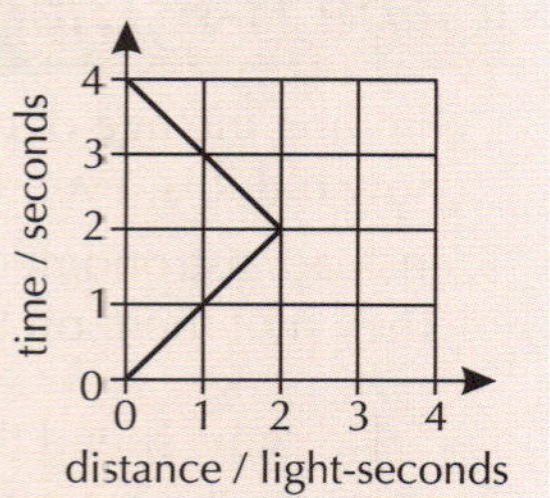

6) You can also use radar to find the **average velocity** of an object **relative** to the Earth. You send **two pulses** separated by a certain **time interval**, to give two separate measurements of the object's **distance**. The difference between the distances shows **how far** the object has moved relative to Earth in the time interval — and **speed = distance ÷ time**.
7) Like most things in physics, this method is based on a couple of **assumptions**:

   1) The **speed** of the **radio waves** is the **same** on the way **to the object** and the **way back** to the telescope.
   2) The **time** taken for the **radio waves** to **reach the object** is the **same** as the **time** taken to **return**.

   For these to be true, the **speed of light** must be **constant**, even though the observer and object are both **moving**, and the **object's speed** must be **much less** than the **speed of light**, so that there are no **relativistic effects** (p.31).
8) More accurate measurements of the speed of distant objects can be made using **Doppler shifts**.

The ***Doppler Effect*** *— the* ***Motion*** *of a Wave's* ***Source*** *Affects its* ***Wavelength***

1) Imagine an ambulance driving past you. As it moves **towards you** its siren sounds **higher-pitched**, but as it **moves away**, its **pitch** is **lower**. This change in **frequency** and **wavelength** is called the **Doppler shift**.
2) The frequency and the wavelength **change** because the waves **bunch together** in **front** of the source and **stretch out behind** it.

The **amount** of stretching or bunching together depends on the **velocity** of the **source**.

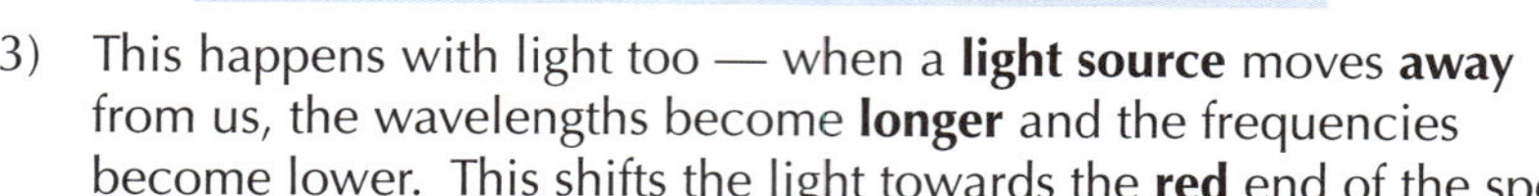

3) This happens with light too — when a **light source** moves **away** from us, the wavelengths become **longer** and the frequencies become lower. This shifts the light towards the **red** end of the spectrum and is called **red shift**.
4) When a light source moves **towards** us, the **opposite** happens and the light undergoes **blue shift**.
5) The amount of red shift or blue shift is determined by the following formula:

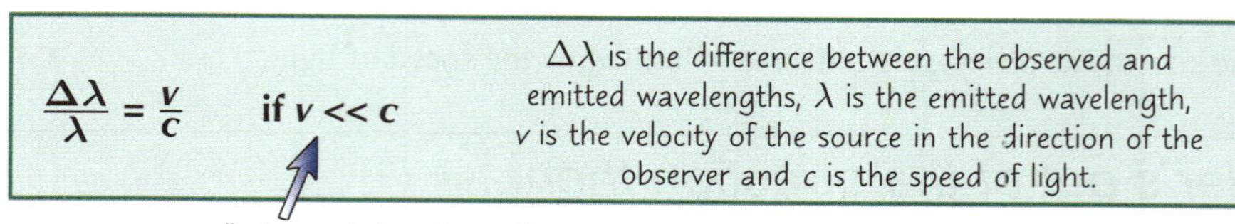

$$\frac{\Delta\lambda}{\lambda} = \frac{v}{c} \quad \text{if } v << c$$

$\Delta\lambda$ is the difference between the observed and emitted wavelengths, λ is the emitted wavelength, v is the velocity of the source in the direction of the observer and c is the speed of light.

$v << c$ means "v is much less than c" (see the next page for why this is important).

Astronomical Distances and Velocities

Stars Produce Absorption Spectra

1) Radiation is emitted from a **very hot** region of a star (called the **photosphere**) in a **continuous spectrum**.
2) **Atoms** in the atmosphere of the star **absorb** certain **wavelengths** of the radiation, producing dark **absorption lines** within the spectrum.
3) Different atoms absorb different parts of the spectrum, resulting in a **characteristic pattern** for each atom. By looking at the absorption lines from a star, and comparing them with known spectra in the lab, the **composition** of the **stellar atmosphere** can be worked out.
4) Once you've worked out which **atoms** make up the **pattern of absorption lines**, you can compare the **position** of the **absorption lines** for **each atom** in the **star's spectrum** with the **same spectrum** recorded in the **lab**.

Stellar spectrum (containing H, He and Na)

Hydrogen
Helium
Sodium
(Lab spectra)

5) This shows **how much** the **spectrum** has been **shifted** by the movement of the star.

Example: The spectrum from this star is **shifted** towards the **red end** of the spectrum, showing that it is **moving away** from Earth.

Stellar spectrum

Lab spectrum

Time Dilation Happens Close to the Speed of Light

1) A big **assumption** of the methods on these two pages is that the **speed** of the object being studied is **much less** than the **speed of light**, ***c***.
2) This matters because **time runs at different speeds** for two objects **moving relative** to each other — but it's only really noticeable **close to** the **speed of light** (which is why you can **ignore** it as long as the object isn't travelling too fast).
3) This effect is called **relativistic time dilation**.

It's not just time that starts doing weird things at high speeds, mass and energy do too (see page 69).

A **stationary** observer measures the time interval between two events as t_0, the **proper time**.
An observer moving at a **constant velocity**, v, will measure a **longer** interval, t, between the two events. t is given by the equation:

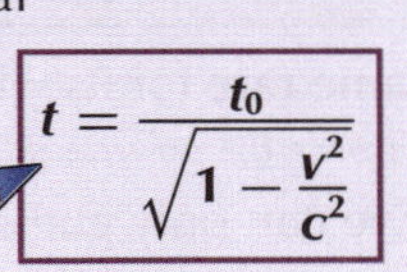

$$t = \frac{t_0}{\sqrt{1 - \frac{v^2}{c^2}}}$$

$\frac{1}{\sqrt{1 - \frac{v^2}{c^2}}}$ is called the **relativistic factor**, γ.

When $v << c$, $\gamma \approx 1$, so the time measured by both observers is almost the same, $t = t_0$ (which is why you can ignore relativistic effects at low speeds).

Practice Questions

Q1 How do we measure the distance to objects in the Solar System using radar?

Q2 What is the Doppler effect? Explain how it can be used to work out the speed of a distant object.

Exam Questions

Q1 A pulse of radio waves is transmitted from a telescope on Earth towards Venus.
The telescope detects the radio waves returning after 4.6 minutes.

a) Show that the distance from Earth to Venus at this time is approximately 4.1×10^{10} m. [3 marks]

b) State two assumptions that your calculation in part a) is based on. [2 marks]

Q2 A spacecraft is travelling through the Solar System, away from Earth. Radar is being used to track its position.
Radar pulses are sent to the spacecraft every second and a receiver detects how long they take to return.

a) Two consecutive pulses take 2.8 s and 3.8 s to return. Estimate the relative velocity of the spacecraft. [3 marks]

b) The estimate won't be accurate as the spacecraft is travelling close to the speed of light.
Calculate the time interval measured between the two pulses by an observer on the spacecraft. [2 marks]

Time dilation occurs close to the speed of light — and in history lessons...

Luckily it only happens close to the speed of light, so as long as 'v << c' is true it's pretty safe to ignore it.

The Hot Big Bang Theory

Right, we're moving on to the BIG picture now — we all like a bit of cosmology...

Hubble Realised that Recessional Velocity is Proportional to Distance

1) The **spectra** from **galaxies** (apart from a few very close ones) all show **red shift** (see page 30). The amount of **red shift** gives the **recessional velocity** — how fast the galaxy is moving away from Earth.
2) When **Hubble** plotted the **recessional velocity** of galaxies against their **distance** from Earth he found that they were **proportional**. This gave rise to **Hubble's law**:

$$v = H_0 d$$

where v = recessional velocity in km s^{-1}, d = distance in Mpc and H_0 = Hubble's constant in km s^{-1} Mpc^{-1}.

A megaparsec (Mpc) is just a unit of distance used in astronomy. 1 Mpc = 3.09×10^{22} m

3) Since distance is very difficult to measure, astronomers used to **disagree** greatly on the value of H_0, with measurements ranging from 50 to 100 km s^{-1} Mpc^{-1}. It's now generally accepted that H_0 lies **between 65 and 80 km s^{-1} Mpc^{-1}** and most agree it's around the **mid to low 70s**. You'll be given a value to use in the exam.

The Universe is Expanding

1) By showing that objects in the universe are **moving away** from each other, Hubble's work is strong evidence that the **universe is expanding**. The rate of expansion depends on the value of H_0, Hubble's constant.
2) The way cosmologists tend to look at this, the galaxies aren't actually moving **through space** away from us. Instead, **space itself** is expanding and the light waves are being **stretched** along with it. This is called **cosmological red shift** to distinguish it from **red shift** produced by sources that **are** moving through space. Don't worry though — you can use the **formula** on page 30 for **both kinds** of red shift.
3) Since the universe is **expanding uniformly** away from **us** it seems as though we're at the **centre** of the universe, but this is an **illusion**. You would observe the **same thing** at **any point** in the universe.

The Age and Observable Size of the Universe Depend on H_0

1) If the universe has been **expanding** at the **same rate** for its whole life, the **age** of the universe is $t = 1/H_0$ (time = distance/speed). This is only an estimate since the universe probably hasn't always been expanding at the same rate.
2) Since no one knows the **exact value** of H_0 we can only guess the universe's age. If H_0 **= 75 km s^{-1} Mpc^{-1} = 2.4×10^{-18} s^{-1}**, then the age of the universe $\approx 1 \div (2.4 \times 10^{-18}\ s^{-1})$ $= 4.1 \times 10^{17}$ s, which is approximately **13 billion years**.
3) The **absolute size** of the universe is **unknown** (and changing) but there is a limit on the size of the **observable universe**. This is simply a **sphere** (with the Earth at its centre) with a **radius** equal to the **maximum distance** that **light** can travel during its **age**. So if H_0 **= 75 km s^{-1} Mpc^{-1}** then this sphere will have a radius of **13 billion light years**. Taking into account the **expansion** of the universe, it is thought to be more like 46-47 billion light years.

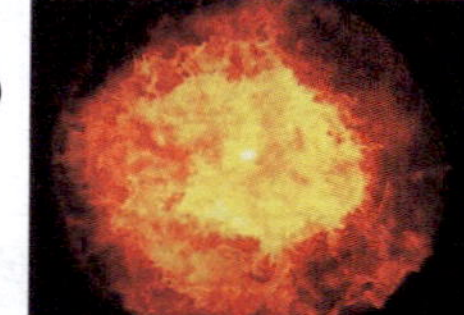
NB — 13 billion candles is too many to put on a birthday cake.

The Red Shift of Galaxies is Strong Evidence for the HBB

1) If the universe is **expanding**, then further back in time it must have been much **smaller**. In fact, if you trace time back **far enough**, and assume that the **expansion** has always been happening, then the entire universe must once have been contained in a **single point** — this is the basis of the **hot big bang (HBB) theory**. The HBB theory is currently the **best explanation** we've got for the state of the universe and is widely **accepted**.

> **THE HOT BIG BANG THEORY (HBB)**: the universe started off **very hot** and **very dense** (perhaps as an **infinitely hot, infinitely dense** singularity) and has been **expanding** ever since.

2) Since Hubble published his findings, the **red shifts** of other, more distant objects in the **universe** have been measured — and they **continue** to fit the **pattern** he predicted, providing more **evidence** for the **HBB theory**. And it doesn't stop there — there's even more evidence for the HBB theory on the next page.

The Hot Big Bang Theory

Cosmic Microwave Background Radiation — More Evidence for the HBB

1) The hot big bang model predicts that loads of **electromagnetic radiation** was produced in the **very early universe**. This radiation should **still** be observed today (it hasn't had anywhere else to go).
2) Because the universe has **expanded**, the wavelengths of this cosmic background radiation have been **stretched** and are now in the **microwave** region.
3) This radiation was **accidentally** detected by Penzias and Wilson in the 1960s.

Properties of the Cosmic Microwave Background Radiation (CMBR)

1) In the late 1980s a satellite called the **Cosmic Background Explorer** (**COBE**) was sent up to have a **detailed look** at the radiation.
2) It found a **continuous spectrum** corresponding to a **temperature** of **2.73 K**.
3) The radiation is largely **isotropic** (the same in all directions) and **homogeneous** (the same everywhere) — it's about the **same intensity** whichever direction you look.
4) There are **very tiny fluctuations** in temperature, which were at the limit of COBE's resolution. These are due to tiny energy-density variations in the early universe, and are needed for the initial '**seeding**' of galaxy formation.
5) The background radiation also shows a **Doppler shift**, indicating the Earth's motion through space. It turns out that our galaxy is rushing towards an unknown mass (the **Great Attractor**) at over a **million miles an hour**.

It's sometimes called 'cosmological microwave background'.

Practice Questions

Q1 What is cosmological red shift? How is it different from ordinary red shift?

Q2 Outline the hot big bang theory.

Q3 Explain why cosmological red shift of light from other galaxies provides evidence for the hot big bang theory for the origin of the universe.

Q4 What is the cosmic microwave background radiation?

Exam Questions

Q1 a) Describe what Hubble's law suggests about the nature of the universe. [2 marks]

b) Using the value $H_0 = 75\ \text{km s}^{-1}\text{Mpc}^{-1}$: (*1 year ≈ 3.16 × 10*^{7} *s, 1 Mpc = 3.09 × 10*^{22} *m*)

i) Calculate H_0 in SI base units. [2 marks]

ii) The age of the universe is approximately equal to H_0^{-1}. Calculate an estimate of the age of the universe in years, and hence the radius of the observable universe, ignoring expansion. [3 marks]

Q2 a) An astronomer observes that the spectral line corresponding to a wavelength of 650 nm has been shifted to 890 nm in the spectrum from a distant galaxy. Estimate the speed at which the galaxy is moving, giving a direction. [2 marks]

b) Use Hubble's law to estimate the distance (in light years) that the galaxy is from us. (Use $H_0 = 2.4 \times 10^{-18}\ \text{s}^{-1}$, $1\ \text{ly} = 9.5 \times 10^{15}\ \text{m}$) [2 marks]

c) Explain why the speed of the galaxy means that your answers to a) and b) are only estimates. [1 mark]

Q3 The cosmic microwave background is a continuous spectrum at a temperature of about 3 K. Explain why its discovery was considered strong evidence for the hot big bang theory for the origin of the universe. [2 marks]

So it's not just about socially awkward physicists trying to get a date then...

The hot big bang model doesn't actually work — not quite, anyway. There are loads of little things that don't quite add up. Modern cosmologists are trying to improve the model using a period of very rapid expansion called inflation.

Temperature and Specific Thermal Capacity

You'd have thought degrees Celsius was a good enough system for anyone — 0 °C = frosty, 25 °C = T-shirts and ice-cream. Apparently not though, as physicists felt the need to invent another one. Physicists, eh...

There's an Absolute Scale of Temperature

1) There is a **lowest possible temperature** called **absolute zero***. Absolute zero is given a value of **zero kelvins**, written **0 K**, on the absolute temperature scale.
2) At **0 K** all particles have the **minimum** possible **kinetic energy** — everything theoretically stops. At higher temperatures, particles have more energy. In fact, with the **Kelvin scale** (named after Lord Kelvin, who first suggested it), a particle's **energy** is **proportional** to its **temperature** (see page 41).
3) A change of **1 K** equals a change of **1 °C**.
4) To change from degrees Celsius into kelvins you **add 273** (or 273.15 to be really specific).

$$K = C + 273$$

5) **Equations** in **thermal physics** use temperatures measured in **kelvins**. (Unless you're dealing with a change in temperature, in which case you can use either kelvins or degrees Celsius — since a change of 1 K equals a change of 1 °C.)

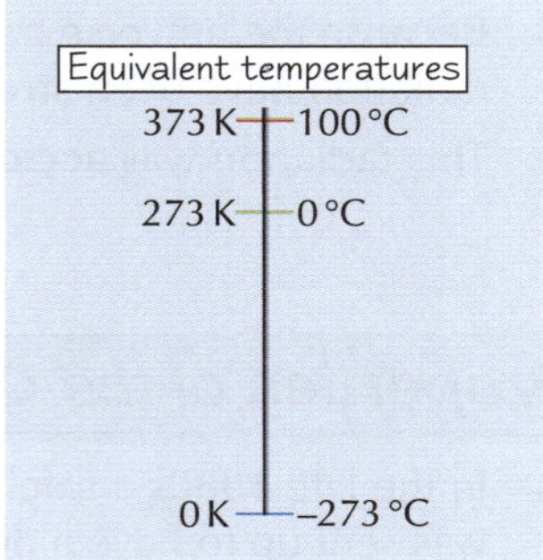

*It's true. –273.15 °C is the lowest temperature theoretically possible. Weird, huh. You'd kinda think there wouldn't be a minimum, but there is.

A Change in Temperature Shows Thermal Energy has been Transferred

Thermal energy is **always** transferred from regions of **higher temperature** to regions of **lower temperature**.

Some energy will also be transferred in the other direction, but overall there'll be a net transfer from hot to cold.

1) Suppose A and B are identical metal blocks. A has been in a **warm oven** and B has come from a **refrigerator**.
2) There is a **net transfer** of **thermal energy** from A to B. As this thermal energy flows out of A, its temperature **decreases**, and as the thermal energy flows into block B, its temperature **increases**. This is because:

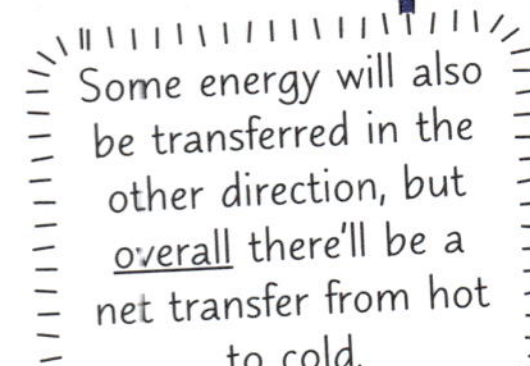

A net **transfer** of **thermal energy** causes a **change** in **temperature**.

3) Once the two blocks reach the **same temperature**, there is no longer a net transfer of thermal energy. We say that the two blocks are in **thermal equilibrium**.

Specific Thermal Capacity is how much Energy it Takes to Heat Something

When you heat something, its particles get more **kinetic energy** and its **temperature** rises.

The **specific thermal capacity** (*c*) of a substance is the amount of **energy** needed to **raise** the **temperature** of **1 kg** of the substance by **1 K** (or 1 °C).

So: **energy change = mass × specific thermal capacity × change in temperature**

Or in symbols: 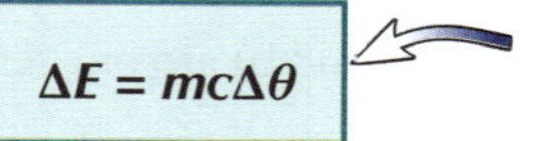

$$\Delta E = mc\Delta\theta$$

Q is sometimes used instead of *E* for the change in thermal energy.

Where ΔE is the energy change in J, *m* is the mass in kg and $\Delta\theta$ is the temperature change in K or °C.

Remember, Δ means 'change in'.

Specific thermal capacity is measured in J kg^{-1} K^{-1} or J kg^{-1} °C^{-1}.

Temperature and Specific Thermal Capacity

You can *Measure* Specific Thermal Capacity in the *Laboratory*

The **method** is the same for **solids** and **liquids**, but the **set-up** is a little bit different:

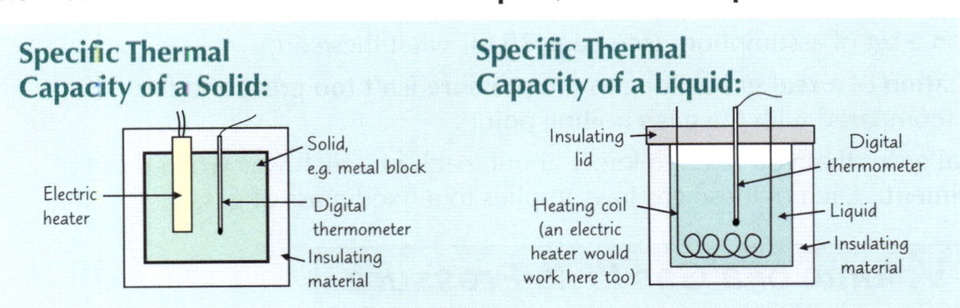

Your value of c will probably be too high by quite a long way, as some of the energy from the heater will get transferred to the air and the container. To minimise this effect, start below and finish above room temperature to cancel out gains and losses. Some energy will also be lost due to resistance in the circuit.

1) **Heat** the substance with the heater so that its temperature increases by about **10 K**.
2) Attach an ammeter and voltmeter to your **electric heater**. You can then calculate the work done by the heater, W, using $W = VIt$ (V is the heater voltage, I is the current and t is the time taken to heat the substance in seconds).
3) If you assume all of the **work done** by the heater is transferred into **thermal energy** in the solid or liquid (so $W = E$), you can then plug your data into: $E = mc\Delta\theta$ to calculate c.

Remember — work is also equal to power × time.

Example: An electric immersion heater is used to heat 0.250 kg of water from 12.1 °C to 22.9 °C. The heater has a voltage of 11.2 V and a current of 5.30 A, and is switched on for 205 s. Calculate the specific thermal capacity of water.

$E = VIt = 11.2 \times 5.30 \times 205 = 12\,168.8$ J $\Delta\theta = 22.9 - 12.1 = 10.8$ °C

$E = mc\Delta\theta$ so $c = \frac{E}{m\Delta\theta} = \frac{12168.8}{0.250 \times 10.8} = 4506.9...$

$= \mathbf{4510\ J\,kg^{-1}\,°C^{-1}}$ (or $4510\ \text{J kg}^{-1}\,\text{K}^{-1}$) **(to 3 s.f.)**

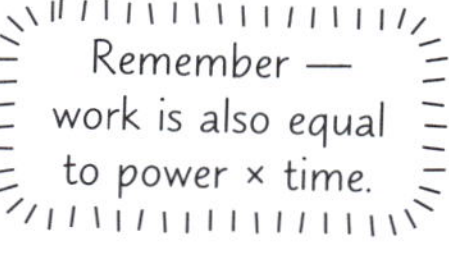

Gary's specific thermal capacity was pretty high...

Practice Questions

Q1 What is meant by absolute zero? Give the value of absolute zero in kelvins and degrees Celsius.

Q2 What is meant by the specific thermal capacity of a substance?

Q3 Briefly describe an experiment to measure the specific thermal capacity of a liquid.

Exam Questions

Q1 a) A stone that has been stored in a freezer is placed in a glass of water at room temperature. Describe how the stone cools the water, in terms of energy transfer. [1 mark]

b) Water has a specific thermal capacity of 4180 J kg^{-1} K^{-1}. Show that the energy change needed to decrease the temperature of 25 g of water from 15°C to 12°C is –310 J. [1 mark]

Q2 A block of tin of mass 0.93 kg is placed in an insulating container then heated by an electric heater. The specific thermal capacity of tin is 210 J kg^{-1} K^{-1}. The heater is connected to a 32 V supply and a current of 5.0 A flows through it when it is switched on.

a) Estimate how long it will take for the temperature of the block of tin to increase from 18.5 °C to 34.2 °C. [2 marks]

b) State how the real time taken for the temperature of the block of tin to increase is likely to differ from your estimate, and explain why. [2 marks]

Specific thermal capacity — it's hot stuff...

You've probably come across a lot of this stuff before, so hopefully it won't be too confusing. You'll be given the equation for specific thermal capacity in the exam, so just make sure you know how to use it.

Ideal Gases

Ideal gases don't actually exist, but they are a really useful tool for understanding how real gases behave.

You Need to Learn the Three **Ideal Gas Laws**

1) An ideal gas is a **model**, based on a set of assumptions (see page 39 for what these are).
2) An ideal gas is a **good approximation** of a **real gas** as long as the **pressure isn't too great** and the **temperature** is **reasonably high** (compared with the gas's boiling point).
3) There are **three gas laws** for ideal gases that you need to know about, each of which was worked out **independently** by **careful experiment**. Each of these gas laws applies to a fixed mass of gas.

Boyle's Law Relates the **Volume** of a Gas to its **Pressure**

Boyle's law states:

At a **constant temperature**, the **pressure** p and **volume** V of a gas are **inversely proportional**.

So: $pV = \text{constant}$

An **ideal gas** obeys Boyle's law at all temperatures.

If you plot pressure against volume for an ideal gas at different temperatures, you'll get a graph like this:

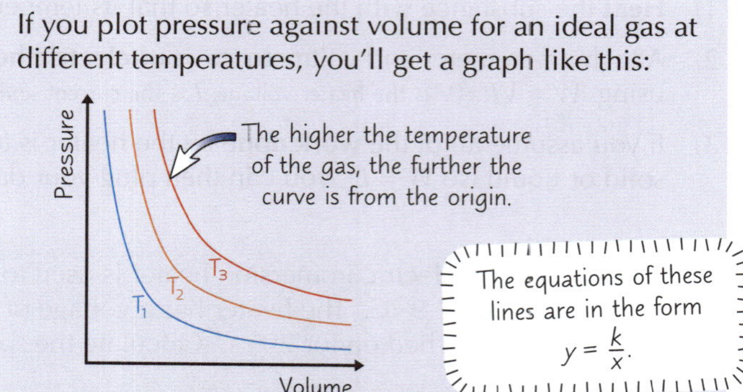

The equations of these lines are in the form $y = \frac{k}{x}$.

You Can **Test Boyle's Law** Using a **Pump** and a **Pressure Gauge**

1) You can investigate the effect of **pressure** on **volume** by setting up the experiment shown on the right.
2) The **oil** confines a parcel of air in a sealed **tube** with **fixed dimensions**. A **tyre pump** is used to **increase** the pressure in the tube and the **Bourdon gauge** records the **pressure**. As the pressure increases the air will **compress** and the volume occupied by air in the tube will **reduce**.
3) Measure the volume of air when the system is at **atmospheric pressure**, then gradually increase the pressure, noting down **both** the pressure and the volume of air. Multiplying them together at any point should give a **constant**.

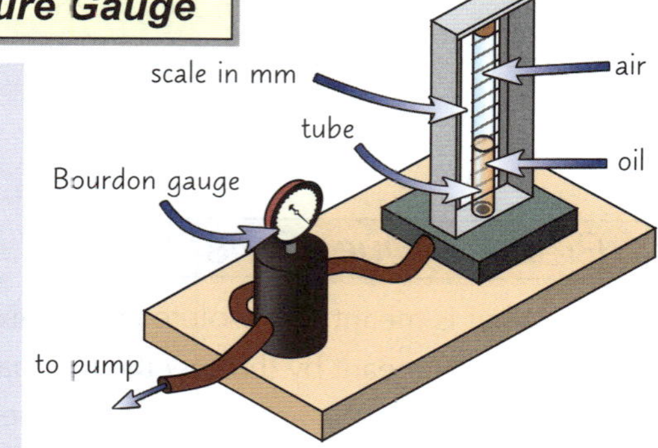

You could also use an electronic pressure sensor, (e.g. a differential pressure monitor) attached to a data logger to record your results automatically.

Charles' Law Relates the **Temperature** of a Gas to its **Volume**

Charles' law states:

At constant **pressure**, the **volume** V of a gas is **directly proportional** to its **absolute temperature** T.

$V/T = \text{constant}$

Ideal gases also obey Charles' law.

If you plot volume against temperature for an ideal gas at a given pressure, you'll get a graph like this:

Volume

–273.15

Temperature (°C)

For any ideal gas the line meets the temperature axis at –273.15 °C — that is, absolute zero. If you used kelvins rather than Celsius, the line would stop at the origin.

Remember, temperature in kelvins is sometimes called absolute temperature.

Ideal Gases

You Can Test Charles' Law Using a Capillary Tube

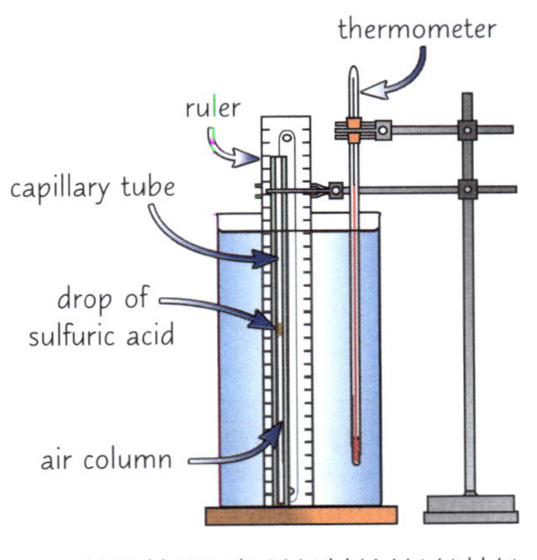

You can use this experiment to estimate the value of absolute zero by continuing (extrapolating) your straight line graph down to where it meets the temperature axis.

1) Set up the experiment shown on the left. You'll need a **capillary tube** containing a drop of **concentrated sulfuric acid** halfway up the tube. The tube should be **sealed** at the bottom, so that a **column of air** is trapped between the bottom of the tube and the acid drop.
2) Fill the beaker with **near-boiling water**. Measure the **length** of the column of air trapped between the bottom of the tube and the drop of acid.
3) As the water cools, regularly record the **temperature** of the water and the **length** of the air column. The length of the trapped air column will **decrease** as the water temperature decreases.
4) **Repeat** with fresh near-boiling water twice more, letting the capillary tube adjust to the new temperature between each repeat. Record the length at the **same temperatures** each time and take an **average** of the three results.
5) Plot your **average results** on a graph of **length** against **temperature** and draw a line of best fit — you should get a **straight line**. This shows that the length of the air column is **proportional** to the temperature.
6) The volume of the column of air is equal to the volume of a cylinder, which is proportional to its length ($V = \pi r^2 l$), so the **volume** is also proportional to the temperature. This agrees with **Charles' law**.

The Pressure Law Relates the Temperature of a Gas to its Pressure

The Pressure law states:

At constant **volume**, the **pressure** p of a gas is **directly proportional** to its **absolute temperature** T.

$$p/T = \text{constant}$$

Unsurprisingly, this is another law that **ideal gases** obey.

If you plot pressure against temperature for an ideal gas at a given volume, you'll get a graph like this:

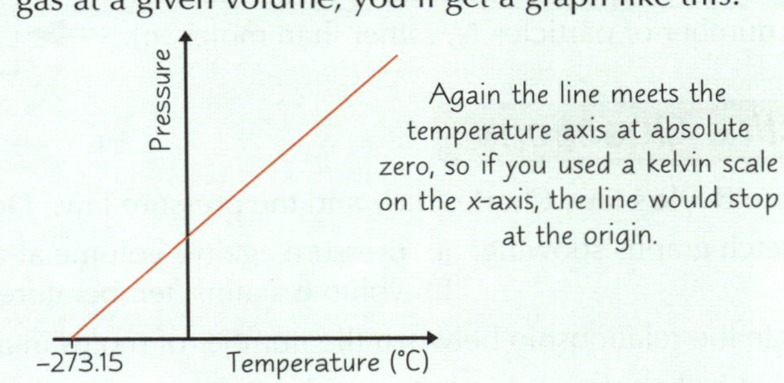

You Can Investigate the Pressure Law Experimentally Too

1) Immerse a stoppered flask of **air** in a beaker of water so that as much as possible of the flask is **submerged**. Connect the flask to a **Bourdon gauge** using a short length of tube — the volume of the tubing must be **much smaller** than the volume of the flask. Record the **temperature** of the water and the **pressure** on the gauge.
2) **Heat** the water for a few minutes then remove the heat source, **stir** the water to ensure it is at a **uniform temperature** and allow some time for the heat to be **transferred** from the water to the air. Record the **pressure** on the gauge and the **temperature**, then heat the water again and **repeat** until the water boils.
3) **Repeat** your experiment **twice** more with **fresh cool water**, taking pressure measurements at the **same set of temperatures** each time.
4) Plot your results on a graph of **pressure** against **temperature**. Draw a **line of best fit** — you'll see it looks like the graph above.
5) This experiment also allows you to **estimate** the value of **absolute zero**, again by **extrapolating** your line of best fit until it reaches the ***x*-axis**.

If the markings on your measuring equipment are quite far apart, you can often interpolate between them (e.g. if the temperature is halfway between the markings for 24 °C and 25 °C you could record it as 24.5 °C). But it's better to use something with a finer scale if you can.

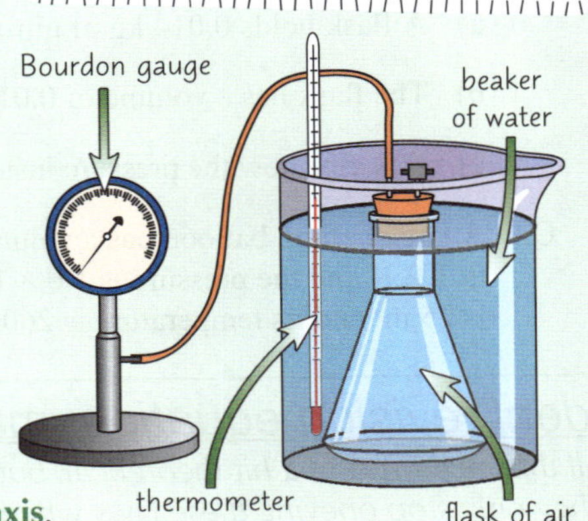

Ideal Gases

If you Combine All Three Laws you get the Ideal Gas Equation

Combining all three gas laws gives the equation: **pV/T = constant**.

1) The constant depends on the amount of gas used. The amount of **gas** can be **measured** in **moles**, ***n*** (see below).
2) The constant is equal to ***nR***, where ***R*** is called the **gas constant**. Plugging this into the equation gives:

$$\frac{pV}{T} = nR$$ or rearranging, $$pV = nRT$$ — **the ideal gas equation**

The value of R is **8.31 J mol^{-1} K^{-1}**. This equation works well (i.e., a real gas approximates to an ideal gas) for gases at **low pressure** and fairly **high temperatures**.

The Boltzmann Constant k is like a Gas Constant for One Particle of Gas

One mole of **any** material contains the same **number of particles**, no matter what the material is. This number is called the **Avogadro constant** and has the symbol N_A. The value of N_A is **6.02 × 10^{23} mol^{-1}**.

1) The **number of particles**, ***N***, in an amount of gas is given by the **number of moles**, ***n***, multiplied by the **Avogadro constant**, N_A. → $N = nN_A$
2) The **Boltzmann constant**, ***k***, is given by $k = \frac{R}{N_A}$. You can think of the Boltzmann constant as the **gas constant** for **one gas particle**, while ***R*** is the gas constant for **one mole of gas**.
3) The value of the Boltzmann constant is **1.38 × 10^{-23} JK^{-1}**.
4) If you combine $N = nN_A$ and $k = \frac{R}{N_A}$ you'll see that: → $Nk = nR$
5) This can be substituted into the ideal gas equation to give this **alternative form** (in terms of number of particles N, rather than moles, n): → $pV = NkT$

The molar mass is the mass of one mole of a substance.

This means that the pressure of a gas is proportional to both the number of particles and the temperature of the gas, and is inversely proportional to its volume.

Practice Questions

Q1 State Boyle's law, Charles' law and the pressure law. Describe how you could test each law experimentally.

Q2 Sketch graphs showing: a) pressure against volume at a constant temperature for an ideal gas.
b) volume against temperature at constant pressure for an ideal gas.

Q3 State the relationship between the number of moles in a gas, n, and the number of particles in a gas, N.

Q4 What is the relationship between the Boltzmann constant, k, and the gas constant, R?

Q5 Write the ideal gas equation in terms of the number of particles in a gas.

Exam Questions

Q1 State the relationship between the pressure and absolute temperature of a gas at a constant volume, and describe two features of a graph of pressure against absolute temperature that show this relation. [3 marks]

Q2 The mass of one mole of nitrogen gas is 0.028 kg.

a) A flask holds 0.014 kg of nitrogen gas. Calculate how many particles of nitrogen gas are in the flask. [2 marks]

b) The flask has a volume of 0.010 m^3 and is at a temperature of 27 °C. Calculate the pressure inside it. [2 marks]

c) Describe how the pressure inside the flask would change if the number of molecules inside was halved. [1 mark]

Q3 A large helium balloon has a volume of 10.0 m^3 at ground level. The temperature of the gas in the balloon is 293 K and the pressure is 1.0 × 10^5 Pa. The balloon is released and rises to a height where its volume is 25 m^3 and its temperature is 260 K. Calculate the pressure inside the balloon at its new height. [2 marks]

Ideal revision equation: marks = (pages read × questions answered)...

All this might sound a bit theoretical, but most gases you'll meet in the everyday world come fairly close to being 'ideal'. They only stop obeying these laws when the pressure's too high or they're getting close to their boiling point.

Kinetic Theory and Internal Energy

***Kinetic theory** tries to **explain** the **gas laws**. It basically models a gas as a series of hard balls that obey Newton's laws. It's a pretty powerful model, and more importantly it's in your exams, so make sure you know about it...*

Lots of Simplifying Assumptions are Used in Kinetic Theory

In **kinetic theory**, physicists picture gas particles moving at **high speed** in **random directions**. To get **equations** like the ones on the last few pages, some **simplifying assumptions** are needed. These are the assumptions that you **need to know**:

1) Particles occupy a **negligible volume** compared with the volume of the container.
2) **Collisions** between particles themselves or at the walls of a container are **perfectly elastic** (so no energy is lost).
3) There are **negligible forces** between particles (except for when they collide).

Kinetic theory also assumes that the gas contains a **large number of particles** that **move rapidly** and **randomly** and that the motion of the particles obeys **Newton's laws** (see below).

A **gas obeying** these **assumptions** is called an **ideal** gas (see p.36-38). Remember — real gases behave like ideal gases as long as the **pressure isn't too big** and the **temperature** is **reasonably high** (compared with their boiling points).

Each Particle in a Gas Goes on a Random Walk

1) There's no way you can **record** the **random motion** of all the particles in a **gas** (without going cross-eyed in the process, of course). Instead you can **model** the movement of the particles by a **random walk**.
2) A **random walk** assumes that each **particle** starts in one place, **moves N steps** in random directions, and ends up **somewhere else**.
3) Here's the path taken by one particle in a box filled with air. The particle **changes direction** each time it **collides** with another particle in the box.

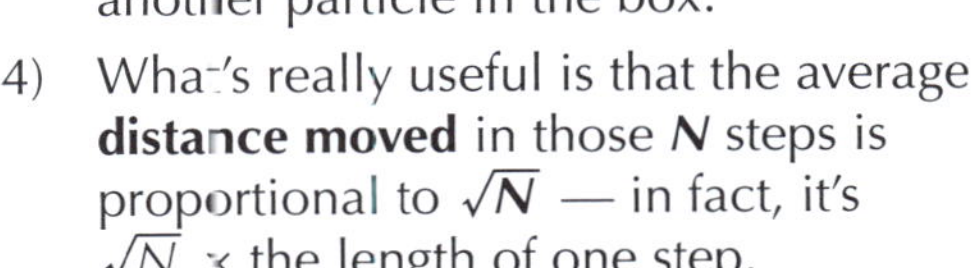

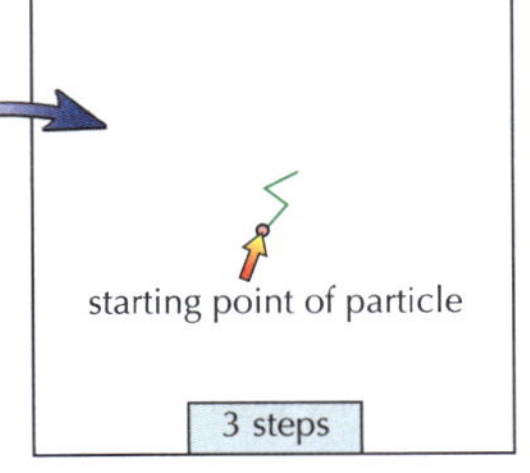

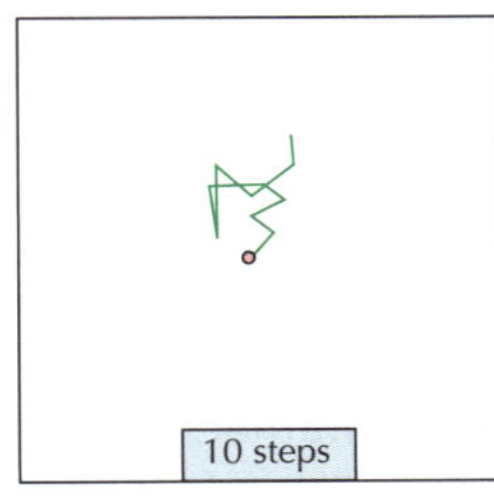

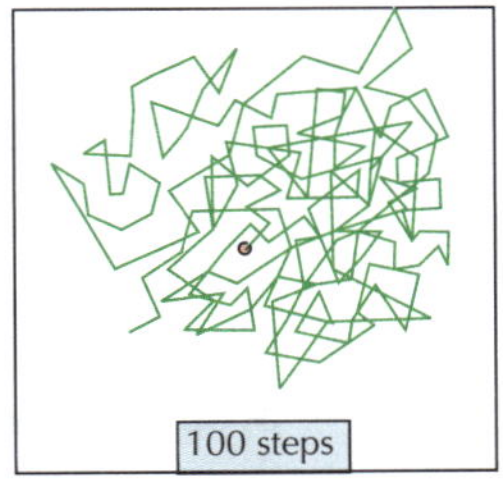

4) What's really useful is that the average **distance moved** in those N steps is proportional to $\sqrt{N}$ — in fact, it's $\sqrt{N}$ × the length of one step.
5) The distance a particle can travel between collisions is usually around 10^{-7} m. So to travel 1 m from its starting point, a particle will have had to take 10 000 000 steps. That's quite a few, so it's no wonder that diffusion is a **slow process**, even if the particles are travelling at high speeds.

Diffusion is the net movement of particles from an area of higher concentration to an area of lower concentration.

Kinetic Theory Assumes Particles Obey Newton's Second Law

1) Kinetic theory assumes that particles in a gas obey Newton's laws.
2) According to Newton's Second Law, the force exerted by a particle in a collision is equal to the rate of change of momentum of the particle, so: $F = \frac{\Delta(mv)}{\Delta t} = \frac{\Delta p}{\Delta t}$
3) This means that if a particle collides with a wall of the container, the faster it is travelling, the more force it exerts on the wall:

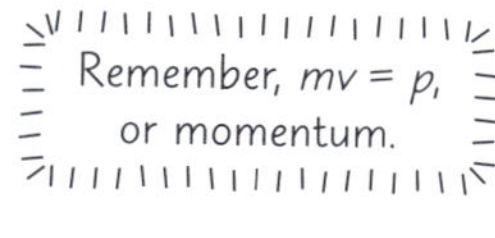
Remember, $mv = p$, or momentum.

A graph of force against time shows how the force acting on (or exerted by) an object changes during an interaction, e.g.:

You know: $F = \frac{\Delta p}{\Delta t}$

so: $F\Delta t = \Delta p$

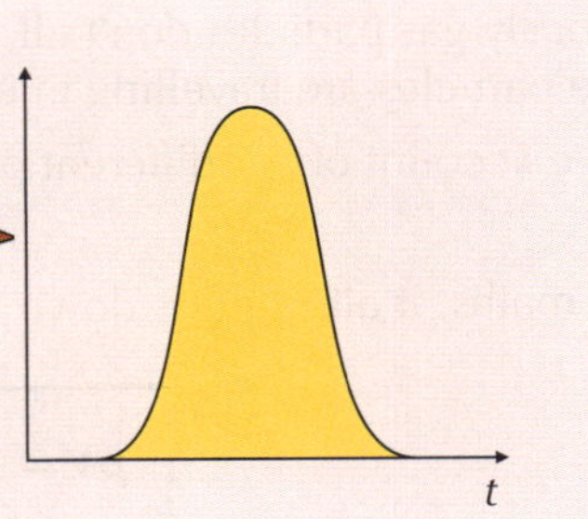

This means the **area under a force-time graph**, $F\Delta t$, is equal to Δp — the **change in momentum**, or the **impulse** of the interaction (you should have met impulse in Year 1).

Kinetic Theory and Internal Energy

The **Root Mean Square** *Speed* **is a Measure of Average Speed**

Before you can combine all of this to make a model of ideal gases, you need to know about the **root mean square speed**:

1) As the particles in a gas are all moving in **different directions**, if you averaged their velocities you'd get **zero**.
2) Instead, you take the average of their **squared velocities**.
   This quantity is called the **mean square speed** and is written $\overline{c^2}$. Its units are m^2s^{-2}.
3) The square root of this number gives you the speed of a typical particle, the **root mean square speed (r.m.s.)**, $\sqrt{\overline{c^2}}$.

To **Understand** *an Ideal Gas, Think About* **Gas Particles** *in a* **Box**

Imagine a cubic box with sides of length ***l*** containing one particle, Q, of mass ***m***.
Say particle Q **moves horizontally** towards **wall A** with velocity ***u***, so its **momentum** is ***mu***. It strikes wall **A**, exerting a **force** on the wall, and heads back in the opposite direction.

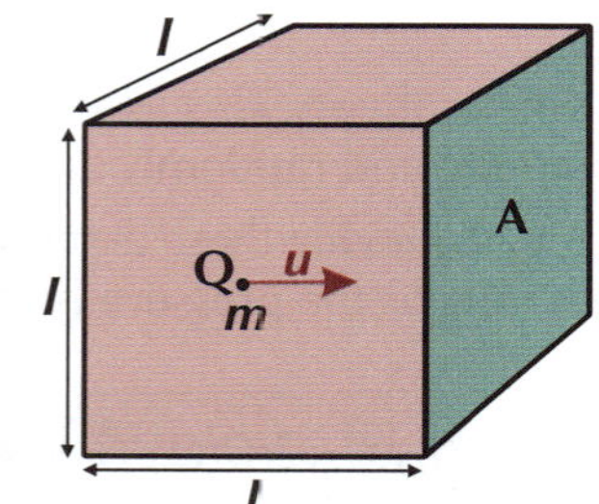

1) Particle Velocity is Proportional to the Pressure

The **faster** the particle, the **larger** its **momentum**, so the **greater the impulse** of the collision and the **larger** the **force** on the wall. The **particle** will also take **less time** to travel across the box and back again, and so will hit the walls more often. And as **pressure = force ÷ area**, the **pressure** will be **greater** too.

Remember: $F\Delta t = \Delta p$, impulse equals change in momentum.

2) The Number of Particles, *N*, is Proportional to the Pressure

Instead of just one particle, imagine you've got a whole stream of them hitting wall A. Each particle exerts a force on the wall as it hits it, so the **total force** on the wall will be **proportional** to the number of particles. And you've guessed it, as **pressure = force ÷ area**, the **pressure** is proportional to the number of particles too.

3) The Volume of the Box is Inversely Proportional to the Pressure

Now imagine you shrink the box. The particles have **less distance** to travel before they hit a wall, so you've **increased** the number of times the particles hit the walls of the box per second, which increases the total force on the wall.
Because the box is now smaller, the **area** of the walls is **smaller**.
So there's a greater force on a smaller area, meaning the **pressure is greater**.

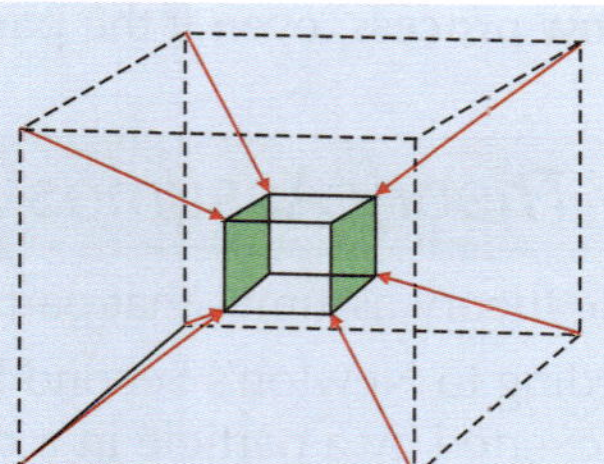

4) Particles Travel in Random Directions at Different Velocities

Obviously gas particles don't all neatly go in the same direction — you can estimate that **a third** of all the particles are travelling in one dimension at any time (as there are three dimensions, *x, y, z*).
To take account of the different particle velocities, you use the **mean square speed**, $\overline{c^2}$ (see above).

If you do the maths, it all whittles down to give you this amazing **equation**:

$$pV = \frac{1}{3}Nm\overline{c^2}$$

Where p is pressure (Pa or Nm^{-2}), V is volume (m^3), N is the number of particles in the gas, m is the mass of a particle (kg) and $\overline{c^2}$ is the mean square speed (m^2s^{-2}).

Kinetic Theory and Internal Energy

Internal Energy is Proportional to Temperature for an Ideal Gas

1) All things (solids, liquids, gases) contain **energy**. The amount of **energy** contained in a system is called its **internal energy**:

Internal energy is the **sum** of the **kinetic** and **potential energy** of the **particles** within a system.

2) All the **internal energy** of an **ideal gas** is due to the **kinetic energy** of its **particles** (there's no potential energy).
3) You can get an expression for the **average kinetic energy of a particle** in an ideal gas in terms of temperature by combining the **ideal gas equation** ($pV = NkT$, page 38), and the equation on p.40 ($pV = \frac{1}{3}Nm\overline{c^2}$):

- $\frac{1}{2}m\overline{c^2}$ is the **average kinetic energy** of an **individual particle** (because $KE = \frac{1}{2}mv^2$).
- If you **equate** the equations in point 3) above, and rearrange a little, you get: (To get the **internal energy** of a whole gas, just multiply this equation by N, the **number of particles** it contains).

$$\textbf{average energy per particle} = \frac{1}{2}m\overline{c^2} = \frac{3}{2}kT$$

k is the Boltzmann constant (1.38×10^{-23} JK^{-1}), T is the temperature (in K)

- You can approximate this to:

$$\textbf{average energy per particle} \approx kT$$

You can rearrange this equation to find the r.m.s. speed from the temperature, T.

4) This means the **internal energy** of an ideal gas is **proportional** to its **absolute temperature** (or vice-versa — the temperature of a gas is proportional to the average energy per particle). So...

A **rise** in **absolute temperature increases** the **kinetic energy** of each particle, causing a rise in **internal energy**.

Practice Questions

Q1 State the three main assumptions required by the kinetic theory of ideal gases.

Q2 A gas particle takes a random walk of 10 000 steps, each of length x. How far does it travel from its starting point?

Q3 Explain how can you calculate the impulse of an interaction from a force-time graph.

Q4 Why should you estimate the average speed of particles in a gas using the root mean square speed?

Q5 What is internal energy? What would cause a rise in the internal energy of an ideal gas?

Q6 What is the average energy per particle in an ideal gas approximately equal to?

Exam Questions

Q1 Some helium gas is contained in a flask of volume 7.0×10^{-5} m^3. Each helium atom has a mass of 6.6×10^{-27} kg, and there are 2.0×10^{22} atoms present. The pressure of the gas is 1.0×10^5 Pa.

a) Calculate the root mean square speed of the atoms. [2 marks]

b) The absolute temperature of the gas is doubled. Use ideas about particle velocity, impulse and momentum to explain why the pressure of the gas will have increased. [4 marks]

Q2 Some air freshener is sprayed at one end of a room. The room is 8.0 m long and the temperature is 22 °C.

a) The average freshener molecule moves at 410 ms^{-1}. Calculate how long would it take for a particle to travel a distance equal to the length of the room. [1 mark]

b) The perfume from the air freshener only slowly diffuses from one end of the room to the other. Explain why this takes much longer than suggested by your answer to part a). [2 marks]

c) Calculate the average energy per air freshener particle. [1 mark]

Matter very simple — my foot...

Kinetic theory basically allows you to treat molecules of gas as little ping-pong balls whizzing about, and (hopefully) it means the ideal gas laws make a bit more sense. If you didn't follow the box stuff on page 40, give it another read, and have a go at sketching diagrams for each point to figure out what's going on.

Activation Energy

Welcome to the big bad world of statistical physics — Ludwig Boltzmann's got a lot to answer for...

*The Average Thermal **Energy** of a Particle is Proportional to the **Temperature***

1) Any particle above absolute zero has some **thermal energy**.

The **average thermal energy per particle** is (very roughly) ***kT***.

k is the Boltzmann constant, $k = 1.38 \times 10^{-23}$ J K^{-1}. *T* is the temperature in kelvins. See page 41 for more.

2) This table gives you an idea of the magnitude of the thermal energy at various temperatures:

Temperature / K	Average thermal energy (approx.) — *kT*		
	J (per particle)	J mol^{-1}	eV (per particle)
1	1×10^{-23}	8	9×10^{-5}
300 (room temp.)	4×10^{-21}	2000	0.03
6000 (Sun's surface)	8×10^{-20}	5×10^{4}	0.5

To convert *kT* in J to J mo^{-1}, multiply by the Avogadro constant (6.02×10^{23} mol^{-1}).

To convert *kT* to eV (electronvolts), divide by the magnitude of the charge on the electron (1.60×10^{-19} C).

3) Particles in matter are **held together** by **bonds**. The **energy** needed to break these bonds in a given substance is the **activation energy**, E_A.

4) The ratio E_A/kT is really important. When ***kT*** is **big enough** compared with E_A, the bonds are broken and the matter comes apart.

***Activation Energy** is the Energy Needed to Make Something Happen*

1) For a process like a change of state to happen, particles need to **'climb'** an **energy barrier**.

2) The **activation energy**, E_A, is the **energy needed** to climb that barrier.

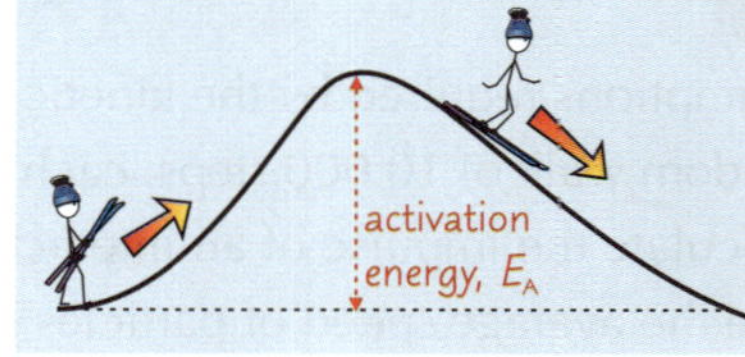

Before you can ski down a mountain, you need to climb to the top of it. So skiing is an activation process.

*Lots of Processes have an **Activation Energy***

1) Lots of processes involving **particles** have activation energies — for example:

a) **A change of state**: the particles need enough energy to break the forces between them.

b) **Thermionic emission**: if you heat up a conductor, electrons are released from the surface. These electrons need enough energy to escape from the attraction of the positive nuclei.

c) **Ionisation in a candle flame**: the molecules in the air need enough energy to split up into individual atoms and ions. This is a similar process to thermionic emission.

d) **Conduction in a semiconductor**: semiconductors will only start to conduct once there are electrons in a high-energy state called the 'conduction band', so electrons need enough energy to jump from the ground state to this higher-energy state.

e) **Viscous flow**: viscous fluids have strong attractive forces between the particles, causing the fluid to 'flow' slowly. As you increase the temperature, you increase the kinetic energy of the particles. This means they have more energy to overcome these forces and so the fluid will be able to flow more easily.

That's why when you've got cold oil in a pan the oil is fairly viscous and doesn't 'run' very easily. As the pan and oil heat up, the oil will flow around the pan much more easily.

2) In each of these examples, the characteristic **activation energy**, E_A, comes from the **random thermal energy** of the particles. You might think, then, that these processes wouldn't happen unless $kT \geq E_A$... but it's not that simple...

Activation Energy

Getting Extra Energy is all about Random Collisions

1) If the **ratio** between the characteristic activation energy and the average energy of the particles (E_A/kT) is too high, nothing happens.
2) As E_A/kT gets down to somewhere around **15–30**, the process starts to happen at a **fair rate**.
3) So some particles must have energies of **15–30 times** the **average energy**.
4) Every time particles **collide**, there's a **chance** that one of them will gain **extra energy** — above and beyond the average kT. If that happens **several times** in a row, a particle can gain energies **much, much higher** than the average.
5) Say f is the fraction of particles with an extra energy E. If E is reasonably big compared to kT, then f will be **small**. (So far so good.)
6) Now to get a particle with an extra energy of $2E$, you need a collision between two particles with an **average extra energy** of about E. So the fraction of particles with an extra energy of $2E$ will be: $f \times f = f^2$

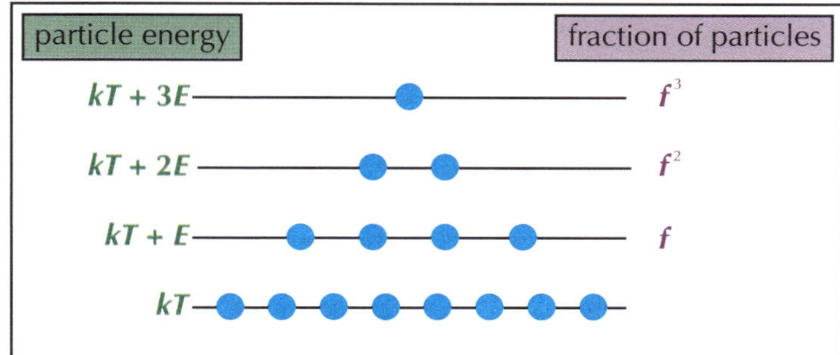

7) You can use the same sort of reasoning to find the fraction of particles with any number of times the extra energy E above the average particle energy. So the fraction of particles with an extra energy of $3E$ is f^3, the fraction of particles with an extra energy of $4E$ is f^4, etc.
8) To end up with an energy of $15kT$ to $30kT$, a particle would have to get **very lucky**, so there will only be a tiny proportion of particles with this energy.
9) Because there are normally **huge numbers** of particles colliding billions of times each second, this small fraction still adds up to a large number of particles.

Edmond's activation energy was 3 cups of coffee and a whiskey chaser.

Practice Questions

Q1 Give an expression for the approximate thermal energy per particle at a given temperature.

Q2 What is activation energy?

Q3 Write down three processes that involve activation energies.

Q4 At what range of values of the ratio E_A/kT does a process begin to occur at a reasonable speed?

Q5 A fraction f of particles have an extra thermal energy E due to collisions with other particles. What fraction of particles have an extra energy equal to $7E$?

Exam Question

Q1 A sample of oil is at 290 K. It is poured into a pan and heated to a temperature of 360 K. *(The Boltzmann constant, $k = 1.38 \times 10^{-23}$ JK^{-1}.)*

a) Estimate the average thermal energy in joules of an oil molecule at:
   i) 290 K [1 mark]
   ii) 360 K [1 mark]

b) Explain in terms of activation energies why the oil is less viscous at 360 K than at 290 K. [3 marks]

Billions of collisions? You'd think they'd look where they're going...

It's like the particle version of Goldilocks — if the ratio between the activation energy and the average energy is too high, nothing happens. You need it just right... or bears will mock you for not knowing about activation energies.

The Boltzmann Factor

I know what you're thinking — if only there was some way to work out the ratio of particles in different energy states. Well, today's your lucky day...

The **Boltzmann Factor** tells you the **Ratio** of Particles in two Energy States

1) You can say particles with different energies are in different quantum **energy states**.
   A particle with an energy of $kT + E$ is in a higher energy state than a particle with an energy of kT.
2) You can find the ratio of particles in two different energy states:

> The **Boltzmann factor**, $e^{-\frac{E}{kT}}$, gives the **ratio** of the **numbers of particles** in energy states E joules apart, at a temperature T in kelvins.

3) The Boltzmann factor is also the **approximate probability** that a particle has an energy of at least E.
4) For an activation energy of E_A, processes start happening **quickly** when E_A/kT is between 15 and 30, so try these values in the **Boltzmann factor**.
5) For $E_A/kT = 15$ the Boltzmann factor is $\sim 10^{-7}$, and for $E_A/kT = 30$ it's only $\sim 10^{-13}$.
6) That means that only about **one in 10^{13}** to **one in 10^7** particles have **enough energy** to overcome the activation energy.
7) But think about a reaction between two gases: gas particles collide about **10^9 times every second**. Every time there's a collision, there's an 'attempt' at the reaction, so even with **so few** particles having enough energy, the reaction can happen in a matter of **seconds**.

Bob and Rodger were both enjoying being in a low energy state.

The **Boltzmann Factor** varies with **Temperature**

For any particular **reaction**, the values of E_A (activation energy) and k (the Boltzmann constant) are **fixed**.
This means that the **only** thing that will change the **Boltzmann factor** is the **temperature**.

If you plot a **graph** of the **Boltzmann factor** against **temperature** you get an s-shaped curve like the ones below.

A graph of the Boltzmann factor against the average thermal energy of the particles would have the same shape, since thermal energy ∝ T.

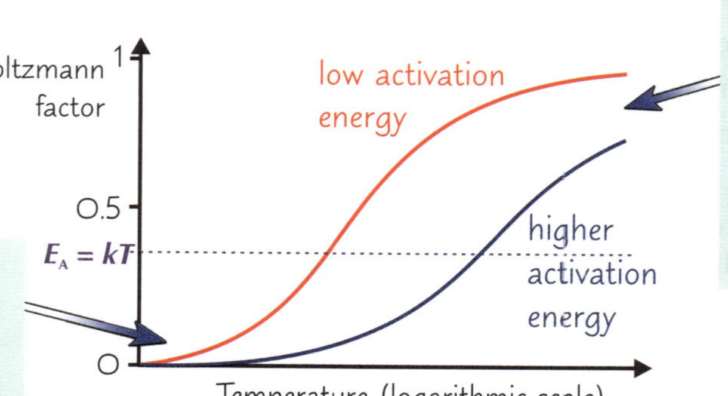

This **shape** shows that at **low temperatures**, the **Boltzmann factor** is also very **low**, so **very few** (if any) particles will have sufficient **energy** to **react** and the reaction will be really **slow**.

At **high temperatures**, the **Boltzmann factor approaches 1**, so nearly **all** the **particles** will have enough **energy** to **react** and the **reaction** will be really **fast**.

In between, the Boltzmann factor **increases rapidly** with **temperature**. So a **small increase** in **temperature** can make a **big difference** to the rate.

The Boltzmann factor and **rate** of a reaction both vary with **temperature**, and it is a **reasonable approximation** to say:

> The **rate** of a reaction with **activation energy** E_A is proportional to the **Boltzmann factor**, $e^{-\frac{E_A}{kT}}$.

The Boltzmann Factor

You can use the Boltzmann Factor to Describe the **Rate of a Reaction**

So, now you can use all that lovely knowledge to answer **exam questions** about the rate of a reaction, just like this one.

Example:

Ben has a flask filled with liquid X. The energy, E_A, binding one molecule in liquid X is 0.40 eV.
a) Estimate the average energy of the liquid molecules at 75 K. ($k = 1.38 \times 10^{-23}$ JK^{-1})

Approx. average energy = $kT = 1.38 \times 10^{-23} \times 75 = 1.035 \times 10^{-21}$ = **1.0×10^{-21} J (to 2 s.f.)**

b) Calculate the ratio E_A/kT for one molecule escaping the liquid at a temperature of 75 K.

$E_A = 0.40$ eV and $kT = 1.38 \times 10^{-23} \times 75 = 1.035 \times 10^{-21}$ J
Convert this energy from J into eV: $(1.035 \times 10^{-21}) \div (1.60 \times 10^{-19}) = 6.468... \times 10^{-3}$ eV
So $E_A/kT = 0.40 \div (6.468... \times 10^{-3}) = 61.835...$ = **62 (to 2 s.f.)**

c) Calculate the Boltzmann factor for the liquid molecules at this temperature.

$e^{-\frac{E_A}{kT}} = e^{-61.835...} = 1.396... \times 10^{-27}$ = **1.4×10^{-27} (to 2 s.f.)**

d) Ben says, 'The ratio of E_A/kT is very high at 75 K, so there will be a rapid rate of evaporation of the liquid X at 75 K.'
Do you agree? Use your answers to parts b) and c) to explain your answer.

I disagree. A high ratio of activation energy to average particle energy means the activation energy is a lot higher than the average energy of the liquid molecules. For a good rate of reaction the E_A/kT ratio should be lower (around 15-30). Also, the Boltzmann factor at 75 K is extremely small. The rate of evaporation will be approximately proportional to the Boltzmann factor, so the rate of evaporation of the liquid will be extremely slow.

Practice Questions

Q1 What is the Boltzmann factor?

Q2 Two energy states are separated by an energy of $E = 7.8 \times 10^{-20}$ J.
Show that, at a temperature of 295 K, the ratio of the number of particles in the higher energy state to the number of particles in the lower energy state is approximately 5×10^{-9}.

Q3 Sketch a graph to show the relationship between the Boltzmann factor and temperature for a reaction.

Q4 How is the rate of a reaction related to the Boltzmann factor?
Describe the rate of reaction if the Boltzmann factor is almost 1.

Exam Question

Q1 Two liquids, A and B, are mixed together at a temperature of 298 K, and begin reacting with one another. ($k = 1.38 \times 10^{-23}$ JK^{-1})

a) Calculate the approximate average energy of one of the molecules in the mixture at this temperature. [1 mark]

b) For a reaction to occur between a molecule of A and a molecule of B, each molecule needs 9.4×10^{-20} J.
Estimate the proportion of molecules that have sufficient energy to react when the mixture is at 298 K. [2 marks]

c) A Bunsen burner is used to increase the temperature of the mixture to 313 K.
Using your answer to part b), estimate the factor by which the rate of reaction increases. [3 marks]

The Boltzmann Factor — not as much fun as the X Factor...

You can think of the Boltzmann factor as a tool in finding the probability of a particle having a certain energy, or the fraction of particles that have that energy. Or you could think of it as a big pair of pants with pink polka dots — up to you.

Magnetic Fields

Magnetic fields — making pretty patterns with iron filings before spending an age trying to pick them off the magnet.

A Magnetic Field is a Region Where a Force is Exerted on Magnetic Materials

Magnetic fields can be represented by **flux lines** (also called **field lines**). Flux lines go from **north to south poles**.

The **strength** of a magnetic field is represented by how **tightly packed** the lines are — the **closer** together the lines, the **stronger** the field.

Each **flux line** always **joins up** the north and south **poles** in one **continuous** line.

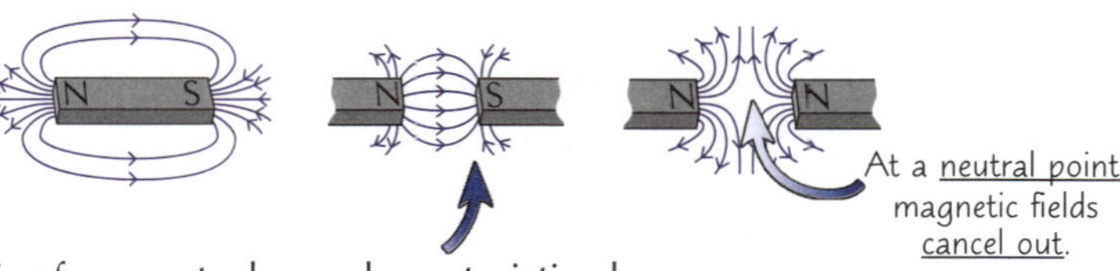

The **flux lines** around a **bar magnet**, or between a pair of magnets, have characteristic shapes.

If the flux lines are **equally spaced** and **in the same direction** the field is **uniform** (i.e. the same everywhere).

There is a Magnetic Field Around a Wire Carrying Electric Current

When **current** flows in a **wire** or any other long straight conductor, a **magnetic field** is induced around the wire.

1) The **field lines** are **concentric circles** centred on the wire.
2) The **direction** of a magnetic **field** around a current-carrying wire can be worked out with the **right-hand rule**.
3) For a **loop** of wire, the field is **doughnut-shaped**, while a **coil** (lots of loops) with **length** forms a **field** like a **bar magnet**.

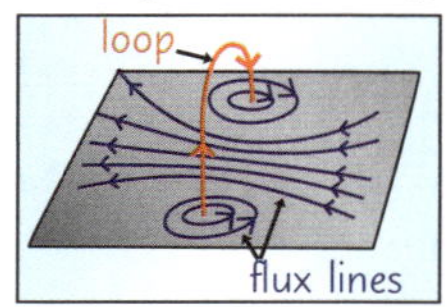

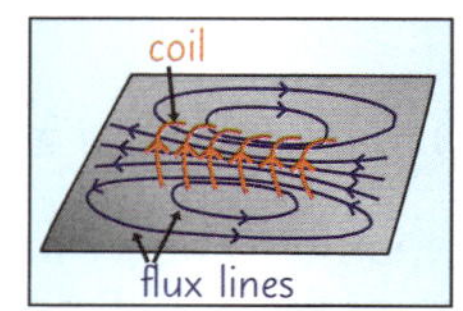

Right-Hand Rule

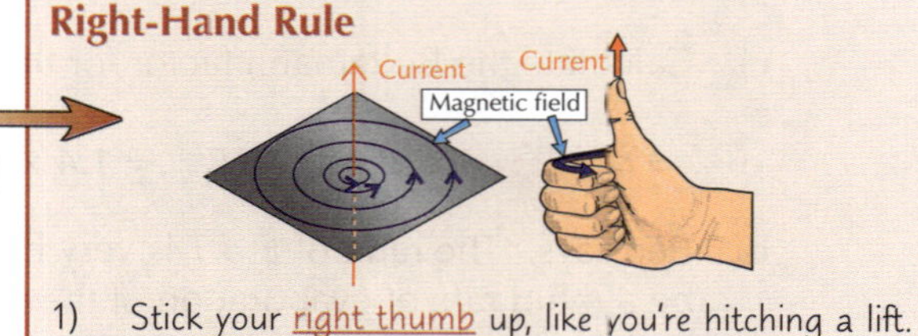

1) Stick your right thumb up, like you're hitching a lift.
2) Your thumb points in the direction of conventional current...
3) ...your curled fingers point in the direction of the field.

A Wire Carrying a Current in a Magnetic Field will Experience a Force

1) If you put a **current-carrying wire** into an **external** magnetic field (e.g. between two magnets), the field around the wire and the field from the magnets **are added together**. The shape of the **resultant flux lines** is a combination of the two fields — here the resultant flux lines are **stretched** around the wire.
2) Flux lines have a tendency to **contract** (get shorter) and **straighten**, which causes an **electromagnetic force** that **pushes** on the wire.
3) If the current is **parallel** to the flux lines, **no force** acts because the fields are **perpendicular**, so they don't affect each other.
4) The **direction** of the force is always **perpendicular** to both the **current** and the **magnetic field** — given by Fleming's left-hand rule:

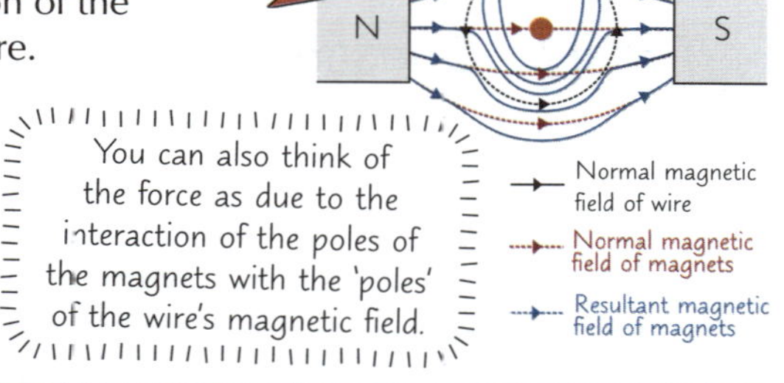

You can also think of the force as due to the interaction of the poles of the magnets with the 'poles' of the wire's magnetic field.

Fleming's Left-Hand Rule

The First finger points in the direction of the external uniform magnetic Field, the seCond finger points in the direction of the conventional Current. Then your thuMb points in the direction of the force (in which Motion takes place).

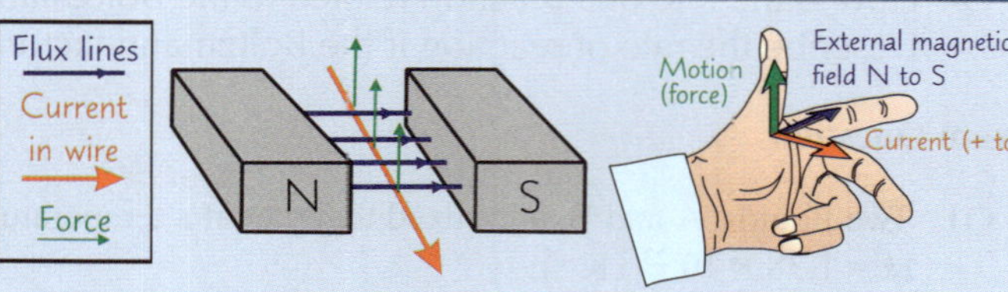

The Force on a Wire is Proportional to the Flux Density

1) The size of the **force**, ***F***, on a current-carrying wire at **right-angles** to a uniform magnetic field is proportional to the **current**, ***I***, the **length of wire** in the field, ***L***, and **magnetic flux density**, ***B***.

$$F = ILB$$

2) **Magnetic flux density**, ***B***, is a measure of the strength of the magnetic field. It is defined as: You might see magnetic fields called ***B*-fields** as a result.

The **force** on **one metre** of wire carrying a **current** of **one amp** at **right angles** to the **magnetic field**.

3) Magnetic flux density is a **vector** quantity with both a **direction** and **magnitude**. It's measured in **teslas**, **T**.

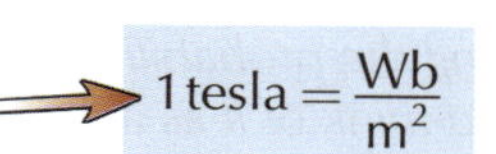

$$1\,\text{tesla} = \frac{\text{Wb}}{\text{m}^2}$$

It helps to think of flux density as the number of flux lines (measured in webers (Wb), see p.48) per unit area.

Magnetic Fields

Use a Digital Balance to Investigate Flux Density

You can use the set-up shown to investigate the **uniform magnetic field** between the poles of a magnet and obtain a value for **flux density**, ***B***. You should use magnets with **poles** on their **largest** faces.

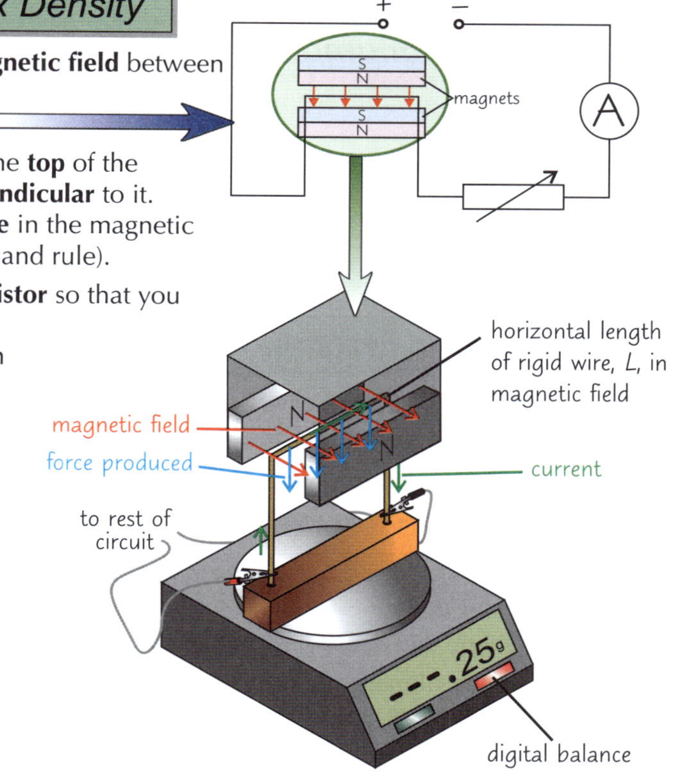

1) A **square hoop** of **rigid** metal wire is positioned so that the **top** of the hoop, **length *L***, passes through the magnetic field, **perpendicular** to it. When a current flows, this horizontal **length of rigid wire** in the magnetic field will experience a downwards **force** (Fleming's left hand rule).
2) The power supply should be connected to a **variable resistor** so that you can **alter** the **current**. Connect the crocodile clips and zero the digital balance when there is **no** current through the wire. Then turn on the power supply.
3) Note the **mass** showing on the digital balance and the **current**. Then use the variable resistor to **change** the **current**. **Repeat** this until you have tested a large range of currents, then do the whole thing **twice** more and calculate the **mean** mass for each current reading.
4) Convert your mass readings into a **force** using $F = mg$. **Plot** the data on a graph of **force *F*** against **current *I***. Draw a line of best fit.
5) Because $\boldsymbol{F = ILB}$, the **gradient** of your graph is equal to $B \times L$. Measure the gradient, then divide by length L to **get a value for *B***.

Practice Questions

Q1 Sketch the lines of magnetic flux around a long straight current-carrying wire and a coil of wire with length. Show the direction of the current and magnetic field on each diagram.

Q2 Write down the equation you would use to find the force on a current-carrying wire at right-angles to a uniform magnetic field.

Q3 A copper bar can roll freely on two copper supports, as shown in the diagram. When current is applied in the direction shown, which way will the bar roll?

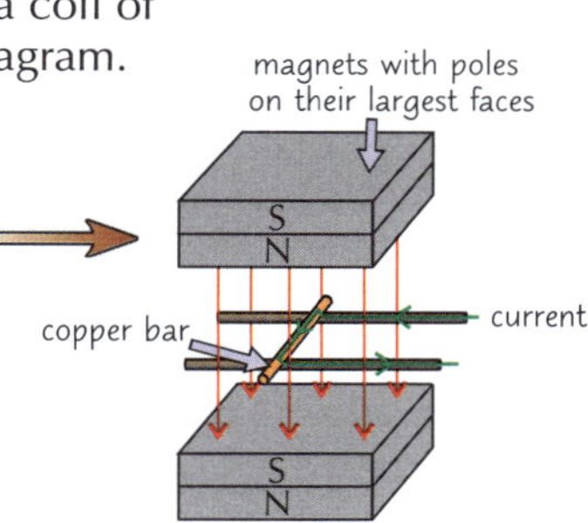

Q4 Describe an experiment you could carry out to determine the uniform magnetic flux density between the poles of a magnet.

Exam Questions

Q1 a) Explain, in terms of flux lines, why a current-carrying wire at right angles to an external uniform magnetic field will experience a force. [2 marks]

b) The flux lines are vertical and go from top to bottom. The current flows from right to left. State the direction in which the force will act. [1 mark]

Q2 A 4.00 cm length of wire carrying a current of 3.0 A runs perpendicular to a uniform magnetic field of strength 2.0×10^{-5} T.

a) Calculate the magnitude of the force on the wire. [2 marks]

b) The wire is rotated so that it runs parallel to the magnetic field. Give the new force on the wire. Explain your answer. [2 marks]

I revised the right-hand rule by the A69 and ended up in Newcastle...

Fleming's left-hand rule is the key to this section — so make sure you know how to use it and understand what it all means. Remember that the direction of the magnetic field is from N to S, and that the current is from +ve to −ve — this is as important as using the correct hand. You need to get those right or it'll all go to pot...

Electromagnetic Induction

Producing electricity by waggling a wire about in a magnetic field sounds like magic — but it's real physics...

Charges Accumulate on a Conductor Moving Through a Magnetic Field

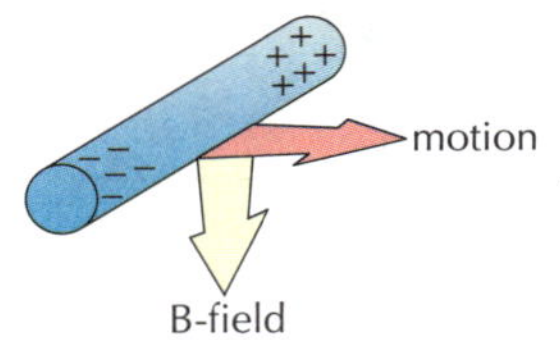

1) If a **conducting rod** moves perpendicular to a magnetic field, the **electrons** in the rod will experience a **force** (see p.46), which causes them to **accumulate** at one end of the rod.
2) This **induces** an **electromotive force (e.m.f.)** across the ends of the rod exactly as connecting a **battery** to it would. If the rod is part of a complete **circuit**, then an induced **current** will **flow** through it too.
3) This process of inducing an e.m.f. is called **electromagnetic induction**.

Changes in Magnetic Flux Induce an Electromotive Force

1) An **e.m.f.** is **induced** whenever there is **relative motion** between a **conductor** and **magnetic flux**.
2) The **conductor** can **move** and the **magnetic field** stay **still** or the **other way round** — you get an e.m.f. either way.
3) An **e.m.f.** is **induced** whenever **lines of flux** are **cut**.
4) **Flux cutting** always induces an e.m.f. but will only **induce** a **current** if the **circuit** is complete.

Think of the Magnetic Flux as the Total Number of Flux Lines in an Area

1) **Magnetic flux density**, ***B***, measures the **strength** of the magnetic field **per unit area**.
2) So, the total **magnetic flux**, ϕ, passing through an **area**, ***A***, perpendicular to a **magnetic field**, ***B***, is defined as:

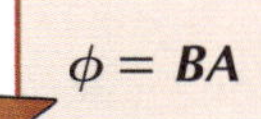

$$\phi = BA$$

(The unit of ϕ is the weber, Wb. 1 tesla = 1 Wb m^{-2}.)

3) When you move a **coil** in a magnetic field, the size of the e.m.f. induced depends on the **magnetic flux** passing through the coil, ϕ, and the **number of turns** on the coil, ***N***. The product of these is called the **flux linkage**, ϕN. The **unit** of flux linkage, ϕN, is the **same** as for flux, ϕ — both are measured in **webers, Wb**.
4) The **size of the e.m.f.** also depends on **how quickly** the flux and the conductor move relative to one another — the **faster** you move a coil in a field, the **greater the size** of the e.m.f. induced.

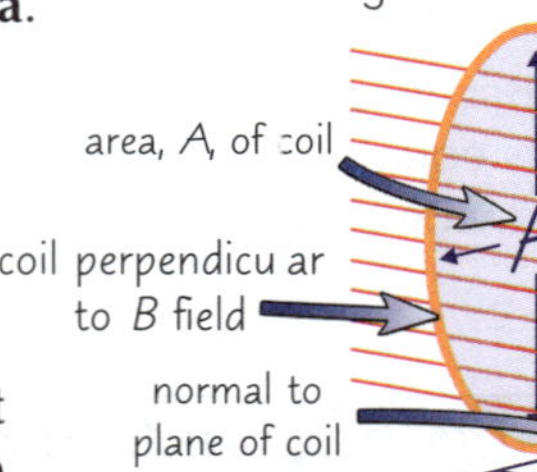

These Results are Summed up by Faraday's Law

FARADAY'S LAW: The **induced e.m.f.** is **directly proportional** to the **rate of change of flux linkage**.

1) **Faraday's law** can be written as:

(The minus sign comes from Lenz's Law — see the next page.)

$$\text{Induced e.m.f., } \varepsilon = -\frac{\text{flux linkage change}}{\text{time taken}} = -\frac{d(\phi N)}{dt}$$

There's more on differential equations like this on page 4.

2) The **size** of the e.m.f. is shown by the **gradient** of a graph of flux linkage (ϕN) against time.
3) The **area under** the graph of e.m.f. against time gives the **flux linkage change**, $\Delta(\phi N)$.

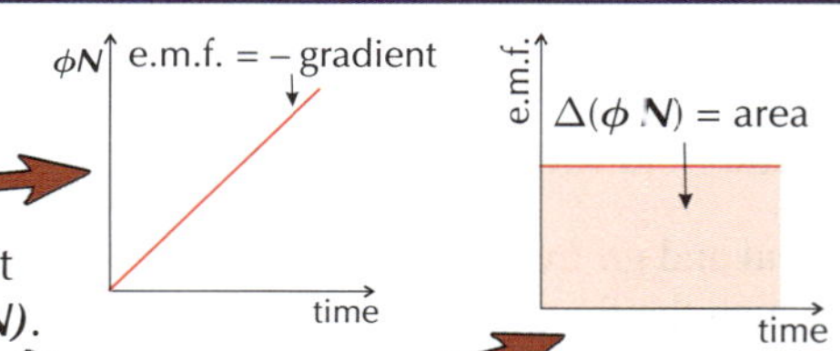

A graph of induced current against time will give the same shape graph as e.m.f. against time (but the area under the graph won't be equal to ϕN).

Example: A conducting rod of length L moves through a perpendicular uniform magnetic field, B, at a constant velocity, v. Show that the e.m.f. induced in the rod is equal to $-BLv$.

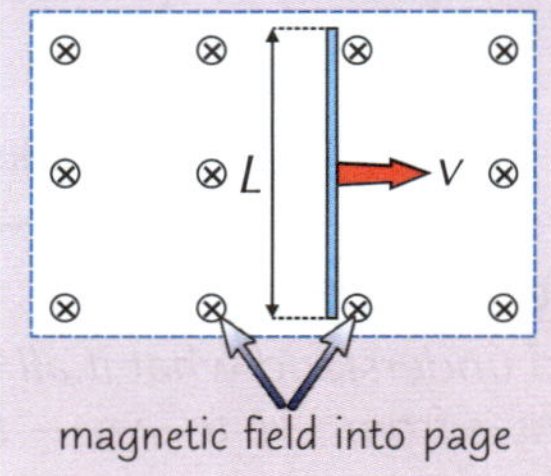

Distance travelled, $s = v\Delta t$ (distance = speed × time)

Area of flux it cuts, $A = Lv\Delta t$

Total magnetic flux cut through, $\phi = BA = BLv\Delta t$

Faraday's (and Lenz's) law gives $\varepsilon = -\frac{d(N\phi)}{dt} = -\frac{d\phi}{dt}$ (since $N = 1$)

So induced e.m.f., $\varepsilon = -\frac{d\phi}{dt} = -\frac{BLvdt}{dt} = -BLv$

You might be asked to find the e.m.f. induced on something more interesting than a rod, e.g. the Earth's magnetic field across the wingspan of a plane. Just think of it as a moving rod and use the equation as usual.

Electromagnetic Induction

The *Direction* of the *Induced E.m.f.* and *Current* are given by *Lenz's Law*

LENZ'S LAW: The **induced e.m.f.** is always in such a **direction** as to **oppose** the **change** that caused it.

That's why there's a minus sign in the equation for induced e.m.f. on the previous page.

1) The idea that an induced e.m.f. will **oppose** the change that caused it agrees with the principle of the **conservation of energy** — the **energy used** to pull a conductor through a magnetic field, against the **resistance** caused by magnetic **attraction**, is what **produces** the **induced current**.
2) **Lenz's law** can be used to find the **direction** of an **induced e.m.f.** and **current** in a conductor travelling at right angles to a magnetic field:

> 1) **Lenz's law** says that the **induced e.m.f.** will produce a force that **opposes** the motion of the conductor — in other words a **resistance**.
> 2) Using **Fleming's left-hand rule** (see p.46), point your thumb in the direction of the force of **resistance** — which is in the **opposite direction** to the motion of the conductor.
> 3) Your **second finger** will now give you the direction of the **induced e.m.f.**
> 4) If the conductor is **connected** as part of a **circuit**, a current will be induced in the **same direction** as the induced e.m.f.

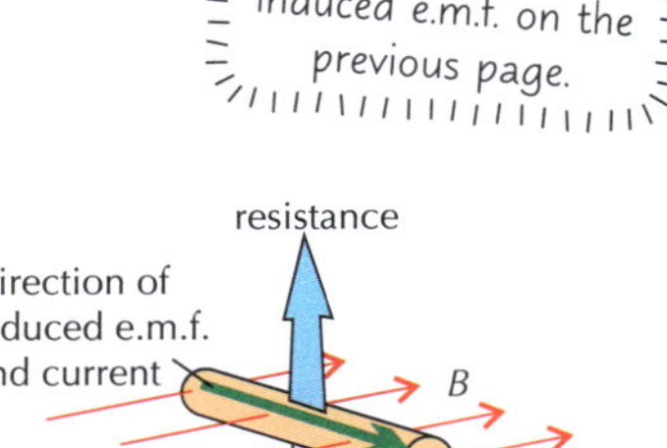

You can *Induce* an *Em.f.* by *Dropping* a *Magnet* Through a *Coil*

You can investigate induced e.m.f. by **dropping** a magnet through a **coil**. An e.m.f. is induced because the conducting coil **cuts** the **flux lines** of the magnet. By connecting a **data logger** or oscilloscope to the coil and recording the e.m.f. in the coil at **very small** time intervals (e.g. 0.002 s), you can plot a graph of induced e.m.f. against time.

> 1) Peak e.m.f. occurs when the **change in flux linkage**, $\Delta(\phi N)$, is **greatest**, which is when each **pole** passes through the coil.
> 2) The **amplitude** of the **second** peak is **greater** because the **speed** of the magnet has increased (so Δt has decreased) and so the rate of change of flux is greater.
> 3) The **area** under each peak is the **same**, because the **total change** in ϕN must be **zero** (there was no e.m.f. before the magnet was dropped, and there is no e.m.f. after).

You'll get **different graphs** if you change the magnet or coil that you use.
If you use a **wider** coil, the magnitude of the induced e.m.f. will be **lower** than if you used a **narrow** coil because the coil will cut **fewer** flux lines.
If you use a **long** bar magnet, there will be a **longer period** between the peaks because there is only a change in ϕN when a pole enters or leaves the coil. The **second** peak will have an even **greater magnitude** and a **shorter duration** because the magnet will have been **accelerating** for **more time** when the second pole passes through the coil, and so will be travelling **faster** (so Δt will be even smaller and $d(\phi N)/dt$ will be even larger).

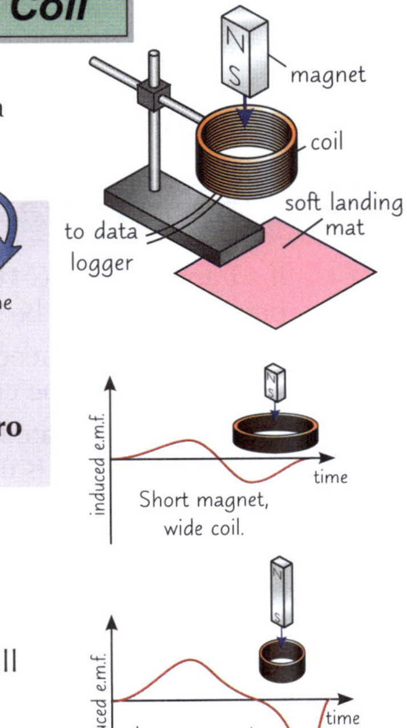

Practice Questions

Q1 What is the difference between magnetic flux density, magnetic flux and magnetic flux linkage?

Q2 Describe and explain the features of a graph of induced e.m.f. against time for a magnet falling through a coil.

Exam Question

Q1 A coil with 150 turns has area 0.23 m^2 and is placed at right angles to a magnetic field of 2.0×10^{-3} T.

a) Calculate the magnetic flux passing through the coil and the magnetic flux linkage in the coil. [2 marks]

b) Over a period of 2.5 seconds the magnetic flux is reduced uniformly to 3.5×10^{-4} Wb. Calculate the e.m.f. induced across the ends of the coil during this time. Explain whether this is positive or negative. [3 marks]

Beware — physics can induce extreme confusion...

OK... I know that might have seemed a bit scary... but the more you go through it, the more it stops being a big scary monster of doom and just becomes another couple of equations you have to remember.

Transformers and Dynamos

Transformers are like voltage aerobics instructors. They say step up, the voltage goes up. They say step down, the voltage goes down. They say star jump, and the voltage does nothing because neither of them are alive — it's just induction.

Transformers *Work by Electromagnetic* ***Induction***

1) **Transformers** are devices that make use of electromagnetic induction to **change** the size of the **voltage** for an **alternating current**. They use the principle of flux linking in two coils of wire, wrapped around an iron core.
2) Alternating current flowing in the **primary coil** produces **magnetic flux**.
3) The changing **magnetic field** passes through the **iron core** to the **secondary coil**, where it **induces** an alternating **voltage** (e.m.f.) of the same frequency as the input voltage.
4) From Faraday's and Lenz's laws (p.48-49), the **induced** e.m.f.s in both the **primary** and **secondary** coils can be calculated:

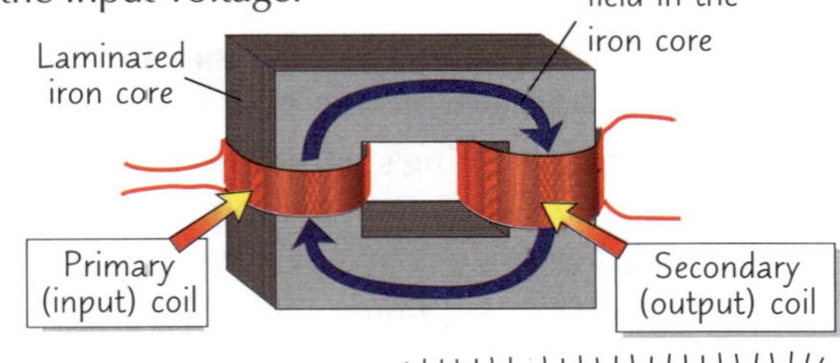

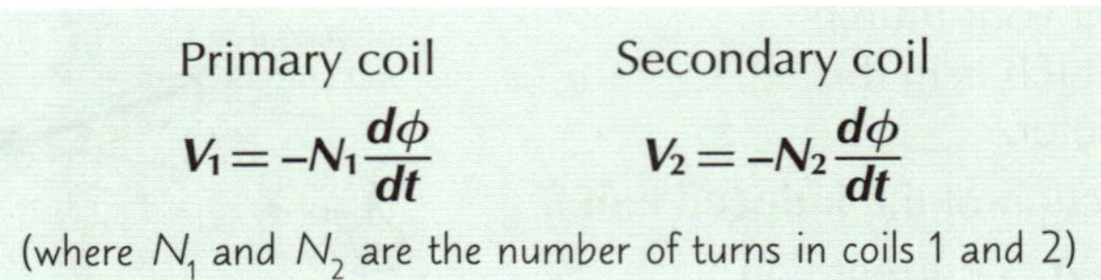

Primary coil $V_1 = -N_1 \frac{d\phi}{dt}$

Secondary coil $V_2 = -N_2 \frac{d\phi}{dt}$

(where N_1 and N_2 are the number of turns in coils 1 and 2)

5) These can be combined to give the equation for an **ideal transformer**: $\frac{V_1}{V_2} = \frac{N_1}{N_2}$ An ideal transformer is one that is **100% efficient**. Unless otherwise specified in a question, you can assume that transformers are ideal.

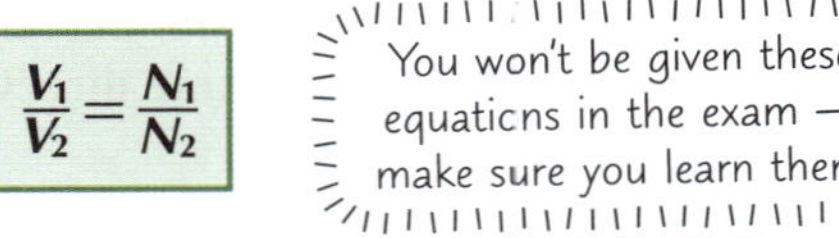

6) For an ideal transformer, power in = power out. Power is current × voltage, so $I_1V_1 = I_2V_2$, so: $\frac{I_2}{I_1} = \frac{N_1}{N_2}$
7) **Step-up** transformers **increase** the **voltage** by having **more turns** on the **secondary** coil than the primary. **Step-down** transformers **reduce** the voltage by having **fewer** turns on the secondary coil.

Transformers are ***Not 100% Efficient***

1) If a transformer was **100% efficient** the **power in** would **equal** the **power out**. However, in practice there will be **small losses** of **power** from the transformer, mostly in the form of **heat**.
2) **Heat** can be produced by **eddy currents** in the transformer's iron core — currents **induced** by the changing magnetic flux in the core. This effect is reduced by **laminating** the core with layers of **insulation**.
3) Heat is also generated by **resistance** in the coils — to minimise this, **thick copper wire** is used, which has a **low resistance**.

Permeability *and* ***Conductivity*** *Affect* ***Transformer Dimensions***

1) Magnetic flux lines are always **continuous** and form a **closed loop**, so you can think of the current-carrying coil as the "**power supply**" of a **magnetic circuit** (although nothing actually flows anywhere). The number of **current turns**, *NI*, is equivalent to the "**e.m.f.**", and the **magnetic flux** is equivalent to the "**current**".
2) The **permeance** of an object is like its 'conductance' — it's the **amount of flux induced** in it for a given number of current turns that surround it. The **higher** the **permeance** of an object, the **greater** the **amount of flux** induced.

permeance, $\Lambda = \frac{\mu A}{L}$

Where A is the cross-sectional area, L is the length and μ is the permeability of the material.

3) Both permeance and conductance are **inversely proportional** to the **length** of the object, and **proportional** to the **cross-sectional area**.
4) When you're **designing** a transformer, you want to make the **permeance** of the core as **high** as possible to get the maximum flux induced in it. Ideally you want the core to be **short** (**low L**) and **fat** (**high A**) and made from a **high permeability** material like **iron**.

The permeability of a material, μ, is the permeance per unit cross-section of a unit length of material.

5) Unfortunately, you also want the **conductance** of the **copper coils** used on a transformer to be as **high** as possible — to limit **energy loss**. So you want to make the right **number of turns** with the **shortest** piece of wire possible, i.e. use small-radius (tight) coils. This doesn't really work when you have a fat core to wrap them around, so you have to try to get a **balance** in **dimensions** to get the **best** overall transformer performance.
6) Unlike an electric circuit, a magnetic circuit will still work if there's an **air** (or vacuum) **gap** in it. So in the case of transformers, if there's an air gap in an otherwise iron core, magnetic flux stills 'flows' around the magnetic circuit. But because air has a very **low permeability** compared to iron, the total amount of flux in the magnetic circuit would be **dramatically lower** than without the air gap.

Transformers and Dynamos

You can Investigate the Turns, Voltage and Current in a Transformer

To investigate the relationship between **number of turns** and the **voltages** across the coils:

1) Set up the equipment as shown. Put two C-cores together and wrap wire around each to make the coils. Begin with 5 turns in the primary coil and 10 in the secondary coil (a **ratio** of 1:2).
2) Turn on the a.c. supply to the primary coil. **Use a low voltage** — remember transformers **increase voltage**, so make sure you keep it at a safe level. Record the voltage across each coil.
3) Keeping V_1 the same so it's a fair test, repeat the experiment with different ratios of turns. Try 1:1 and 2:1. Divide N_2 by N_1 and V_2 by V_1. You should find that for each ratio of turns, $\frac{N_2}{N_1} = \frac{V_2}{V_1}$.

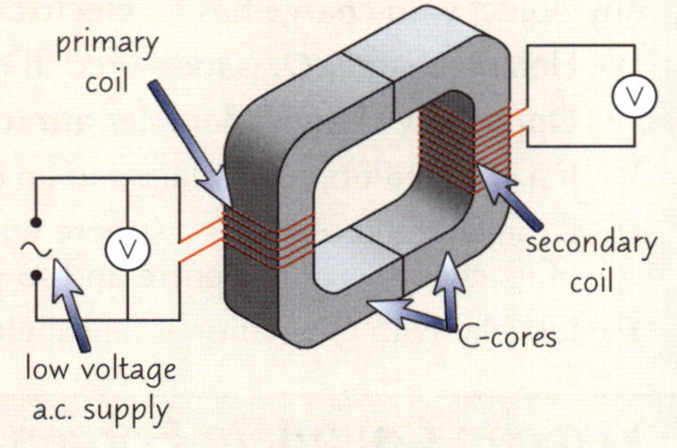

To investigate the relationship between **number of turns**, **voltage** across and **current** of the transformer coils:

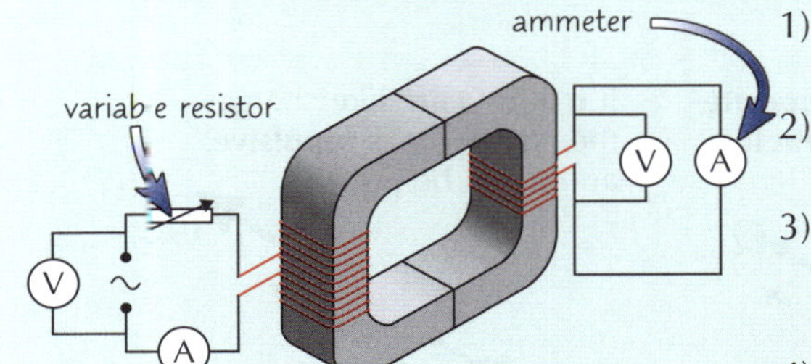

1) Use the same equipment as above, but add a **variable resistor** to the primary coil circuit and an **ammeter** to both circuits.
2) Turn on the power supply and **record** the **current through** and **voltage across** each coil.
3) Leaving the number of turns **constant**, adjust the variable resistor to change the input current. Record the current and voltage for each coil, then **repeat** this process for a **range** of input currents.
4) You should find that for each current, $\frac{N_2}{N_1} = \frac{V_2}{V_1} = \frac{I_1}{I_2}$.

Dynamos Convert Kinetic Energy into Electrical Energy

1) We've seen that **relative motion** of a **coil** and **flux** results in a change in **flux linkage** and an **e.m.f.** is induced in the coil (p.48).
2) **Dynamos**, or generators, **induce** an electric **current** by **rotating** a **coil** in a magnetic field.
3) The output **voltage** and **current** change direction with every **half rotation** of the coil, producing **alternating current (a.c.)**.
4) A **split ring commutator** is used to change this **a.c. current** to **direct current (d.c.)**. This current is carried to an external circuit using **brushes**.

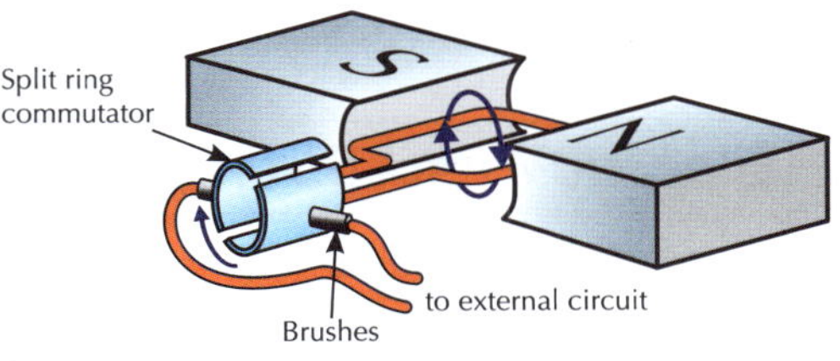

Practice Questions

Q1 Explain, in terms of flux, how a transformer works. How could you reduce losses in the core due to eddy currents?

Q2 What is meant by the permeance of an object? What is it equivalent to in an electrical circuit?

Q3 What is the effect on the permeance of an iron core transformer of making it:
a) fatter, b) shorter, c) have an air gap?

Q4 How could you demonstrate the relationship between the number of turns and the voltages across transformer coils?

Q5 Describe what a dynamo is, and explain how one works.

Exam Question

Q1 An ideal transformer with 150 turns in the primary coil has an input voltage of 9.0 V.
a) State what is meant by an 'ideal transformer'. [1 mark]
b) Calculate the number of turns needed in the secondary coil to step up the voltage to 45 V. [2 marks]
c) The input current for the transformer is 1.5 A. Calculate the output current of the transformer. [2 marks]

Breathe a sigh of relief and pat yourself on the back...

...well done, you've reached the end of the section. Don't let all those equations get you down — once you've learnt them and can use them blindfolded, even the trickiest looking exam question will be a walk in the park...

Electric Fields

*Unlike gravitational fields (p.24), electric fields can be attractive or repulsive. It's all to do with **charge**.*

There is an Electric Field around a Charged Object

Any object with **charge** has an **electric field** around it — the region where it can attract or repel other charges.

1) Electric charge, Q, is measured in **coulombs** (C) and can be either positive or negative.
2) **Oppositely** charged particles **attract** each other. **Like** charges **repel**.
3) If a **charged object** is placed in an electric field, then it will experience a **force**.
4) If the charged object is a **sphere** with **evenly distributed** charge (**spherically symmetrical**), it will act as if all of its **charge** is at its **centre** and so you can model it as a **point charge** (at distances larger than its radius).
5) Just like with gravitational fields, **electric fields** can be represented by **field lines**.

You can Calculate Forces using Coulomb's Law

You'll need **Coulomb's law** to work out F — the force of attraction or repulsion between two point charges (or two objects that behave as point charges).

COULOMB'S LAW:

$$F_{electric} = \frac{kqQ}{r^2} \quad \text{where } k = \frac{1}{4\pi\varepsilon_0}$$

ε_0 = permittivity of free space, equal to $8.85 \times 10^{-12}\ C^2N^{-1}m^{-2}$.
q and Q are the charges (in C)
r is the distance between q and Q (in m)

If the charges are **opposite** then the force is **attractive**. F will be **negative**.

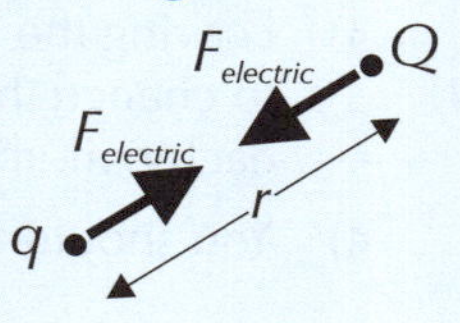

If q and Q are **like** charges then the force is **repulsive**, and F will be **positive**.

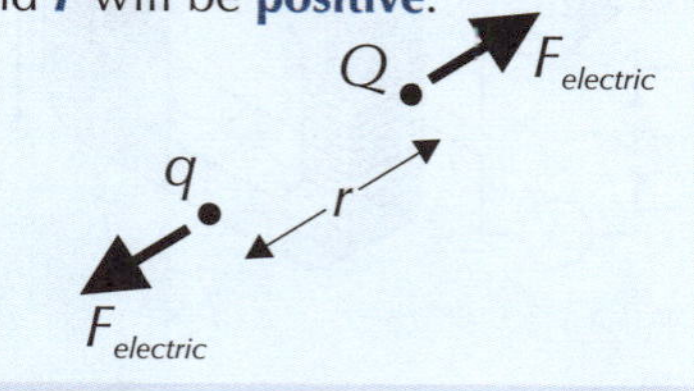

1) The force on q is always **equal** and **opposite** to the force on Q.
2) It's an **inverse square law** (like on p.24). The further apart the charges are, the weaker the force between them.
3) k is called the electric force constant. It's equal to about $9.0 \times 10^9\ Nm^2C^{-2}$.
   You'll be given its value in the exam, either in the data and formulae booklet, or in the question.

Electric Field Strength is Force per Unit Charge

Electric field strength, $E_{electric}$, is defined as the **force per unit positive charge** — the force that a charge of +1 C would experience if it was placed in the electric field.

1) $E_{electric}$ is a **vector** pointing in the **direction** that a **positive charge** would **move**.
2) The units of $E_{electric}$ are **newtons per coulomb** (NC^{-1}).
3) Field strength often depends on **where you are** in the field (see below).
4) A **point charge** — or any body that behaves as if all its charge is concentrated at the centre — has a **radial** field.

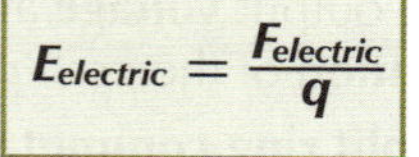

$$E_{electric} = \frac{F_{electric}}{q}$$

$F_{electric}$ is the force on a 'test' charge q.

In a Radial Field, $E_{electric}$ is Inversely Proportional to r^2

1) In a **radial field**, $E_{electric}$ depends on the distance r from the point charge Q (or from the centre of the source of the radial field).

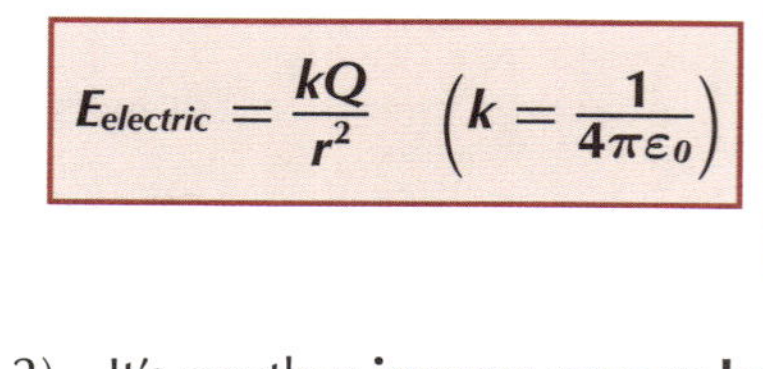

$$E_{electric} = \frac{kQ}{r^2} \quad \left(k = \frac{1}{4\pi\varepsilon_0}\right)$$

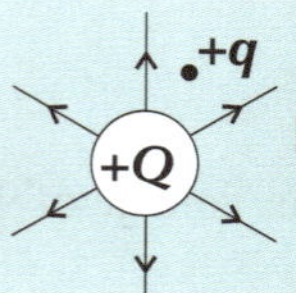

For a **positive** Q, the small positive 'test' charge q would be **repelled**, so the field lines point **away** from Q.

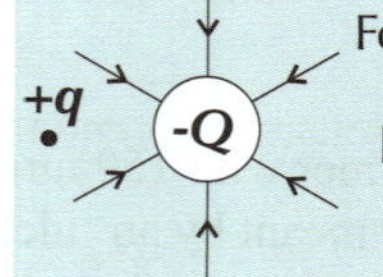

For a **negative** Q, the small positive charge q would be **attracted**, so the field lines point **towards** Q.

2) It's another **inverse square law** — $E_{electric} \propto \frac{1}{r^2}$
3) Field strength **decreases** as you go **further away** from Q — on a diagram, the **field lines** get **further apart**.

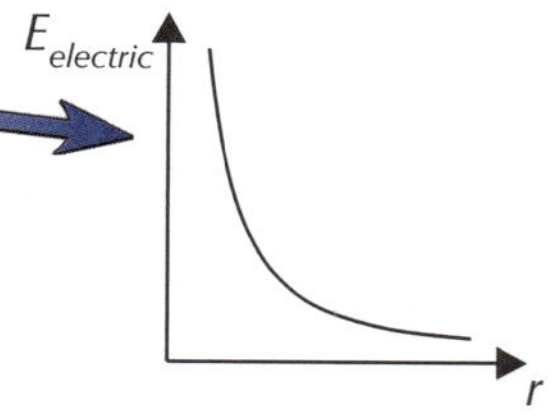

$E_{electric}$ (vertical axis), r (horizontal axis)

Electric field lines always go from + to −. The field around a point charge is spherically symmetrical, so the field lines should be evenly spaced.

Electric Fields

A Charge in an Electric Field has Electric Potential Energy

The electric potential energy is the **work** that would need to be done to move a small charge ***q*** from infinity to a distance ***r*** away from a point charge ***Q***.

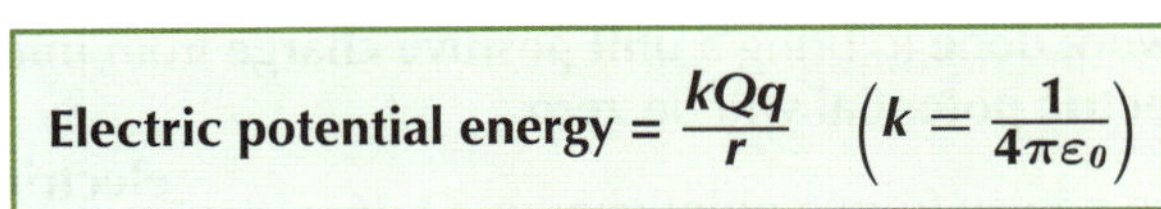

$$\text{Electric potential energy} = \frac{kQq}{r} \quad \left(k = \frac{1}{4\pi\varepsilon_0}\right)$$

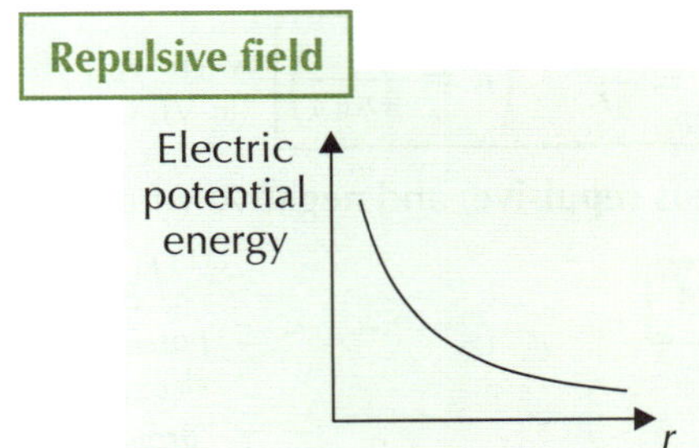

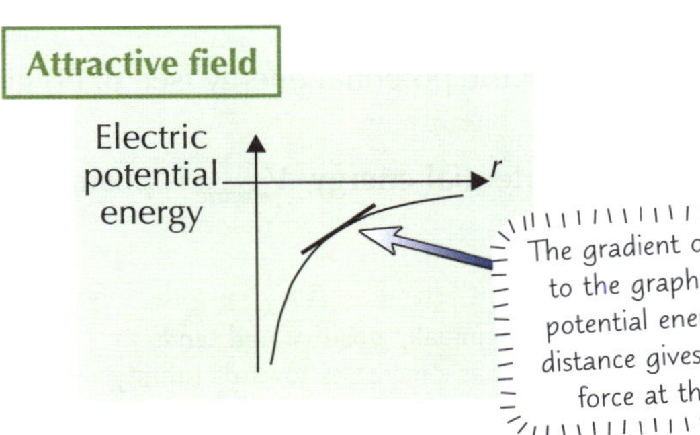

The gradient of a tangent to the graph of electric potential energy against distance gives the electric force at that point.

1) At an **infinite** distance from ***Q***, a charged particle ***q*** would have **zero potential energy**.
2) In a **repulsive** force field (e.g. ***Q*** and ***q*** are both positive) you have to **do work** against the repulsion to bring ***q*** closer to ***Q***. The charge ***q* gains** potential energy as ***r* decreases**.
3) In an **attractive** field (e.g. ***Q*** negative and ***q*** positive) the charge ***q* gains** potential energy as ***r* increases**.

If you **move** a unit charge between two distances, r_1 and r_2, and **change** its **electric potential energy**, you have to apply a **force** and do **work**. The force applied is equal to the electric force. For a point charge ***q*** (and therefore also for a spherically symmetric charge, see p.52) you can plot the **electric force**, $F_{electric}$, against the **distance**, ***r***, from the charge producing the **electric field**, ***Q***.

This is an **inverse square law** (p.24) and the **area** under the curve (p.26) between r_1 and r_2 gives the **change in** electric potential energy.

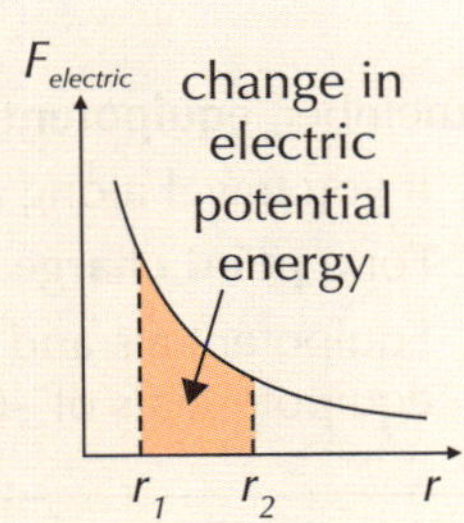

Practice Questions

Q1 Write down Coulomb's law.

Q2 Draw the electric field lines due to a positive charge, and due to a negative charge.

Q3 What is meant by electric field strength? Write down an equation for the electric field strength of any electric field.

Q4 Using your answers to Q2 and Q3, write down a formula for the electric field strength at a distance *r* away from a point charge *Q*.

Q5 Draw a graph of $F_{electric}$ against *r* for the field described by Coulomb's law.
What does the area under the graph between distances r_1 and r_2 represent?

Exam Question

$e = -1.60 \times 10^{-19}$ C, $\varepsilon_0 = 8.85 \times 10^{-12}$ $C^2N^{-1}m^{-2}$ (Fm^{-1}).

Q1 a) Find the electric field strength at a distance of 1.00×10^{-10} m from a proton. [2 marks]

b) A point charge of 6.4×10^{-19} C is placed at a distance of 3.25×10^{-9} m from the proton. Calculate the electric potential energy of this point charge. [2 marks]

c) Sketch a graph showing how the electric potential energy of the point charge depends on the distance, *r*, from the proton. [1 mark]

d) State the quantity that is represented by the gradient of this graph at a specific point. [1 mark]

Electric fields — one way to roast beef...

At least you get a choice with electric fields — positive or negative, attractive or repulsive, chocolate or strawberry...

Electric Potential

Electric potential is all to do with how much energy a charge has based on where it is in an electric field.

Electric Potential is *Potential Energy per Unit Charge*

Electric potential, $V_{electric}$, is electric **potential energy** per **unit positive charge**. The electric potential at a point in an electric field is the **work done** to bring a **unit positive charge** from **infinity** to that point. This means that at **infinity**, the **electric potential** will be **zero**.

For a small charge, q, at a distance r away from a point charge, Q, $V_{electric} = \frac{\text{electric potential energy}}{q}$.

Substituting for electric potential energy (see p.53) gives:

$$V_{electric} = \frac{kQ}{r} \quad \left(k = \frac{1}{4\pi\varepsilon_0}\right)$$

where $V_{electric}$ is electric potential (in V), Q is in C and r is in m.

As with **electric potential energy**, $V_{electric}$ is **positive** when the force is **repulsive**, and **negative** when it is **attractive**.

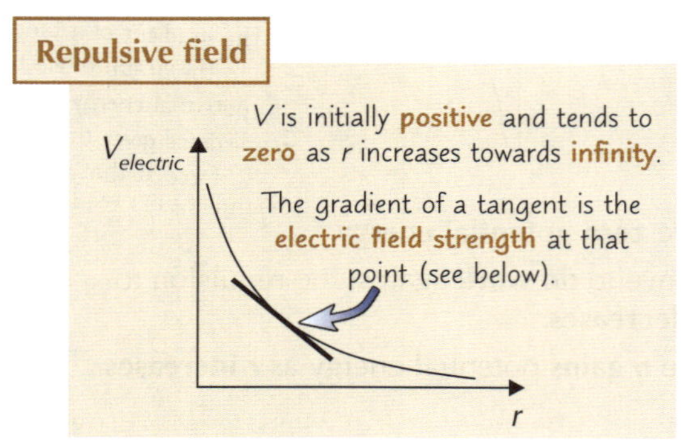

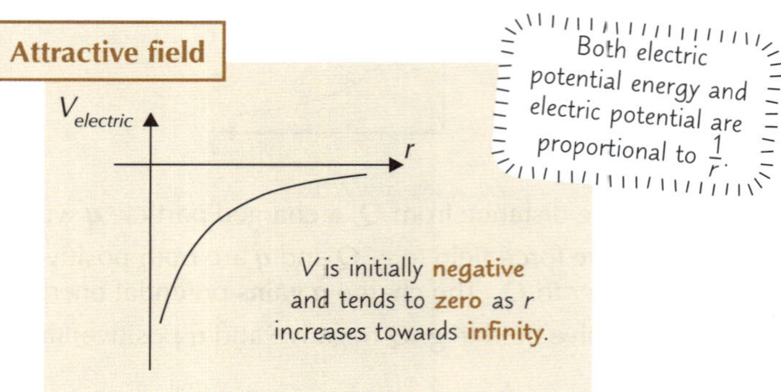

Remember, **equipotentials** show all the points in a field which have the same **potential** (p.27).

1) If you travel along a line of **equipotential** you don't lose or gain energy (no work is done).
2) For a **point charge** or **spherically symmetric charge** the equipotentials are spherical surfaces.
3) Equipotentials and field lines are perpendicular. The diagram shows equipotentials of –60, –50, and –40 V around a point charge Q.

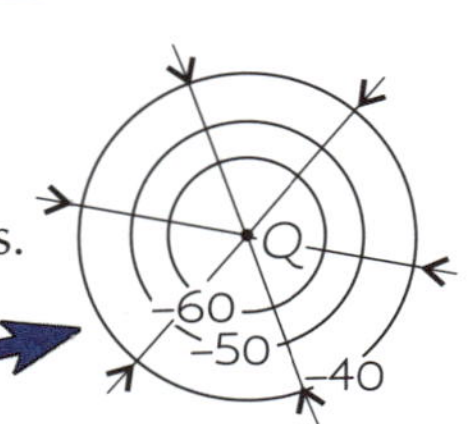

Electric Potential is Linked to *Electric Field Strength*

The **electric field strength, $E_{electric}$** is equal to the negative of the rate of change of electric potential with distance:

$$E_{electric} = -\frac{dV_{electric}}{dr}$$

There's a minus sign because to increase the charge's potential (and potential energy), you have to do work against the force — i.e. they 'act' in opposite directions.

This means that $E_{electric}$ can also be measured in Vm^{-1} and the **gradient** of a graph of $V_{electric}$ **against r** is $E_{electric}$, as you've seen above.

You also need to know that the **area** under a graph of $E_{electric}$ **against r** between two distances gives $\Delta V_{electric}$.

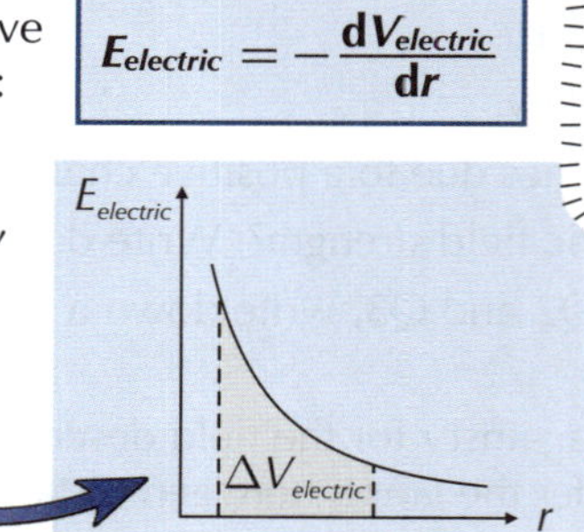

Field Strength is the *Same Everywhere* in a *Uniform Field*

A **uniform field** can be produced by connecting two **parallel plates** to the opposite poles of a battery.

1) Field strength $E_{electric}$ is the **same** at **all points** between the two plates and is given by:

$$E_{electric} = \frac{V}{d}$$

where V is the **potential difference** between the plates and d is the distance between them.

2) This formula comes from the equation above for $E_{electric}$, the rate of change of potential is constant for a uniform field.

+ 400 V
+ 300 V
+ 200 V
+ 100 V
0 V

The **field lines** are **parallel** to each other and evenly spaced.

The **equipotential surfaces** are **parallel** to the **plates**, and **perpendicular** to the **field lines**. They're also evenly spaced and symmetric.

Electric Potential

There are Similarities between Gravitational and Electric Fields...

If a lot of the stuff on the previous couple of pages sounded strangely familiar it could be because it's very similar to the stuff on gravitational fields (or it could be because you've learnt it before — this is a revision book after all). Anyway, there are **four** big **similarities** between **electric** and **gravitational fields** that you need to know — read on.

Gravitational	Electric
Gravitational field strength, g, is **force** per **unit mass**.	Electric field strength, $E_{electric}$, is **force** per **unit positive charge**.
Newton's law of gravitation for the **force** between two point masses is an **inverse square law**.	Coulomb's law for the electric **force** between two point charges is also an **inverse square law**.
The **field lines** for a point mass (or a spherically symmetric mass)...	The **field lines** for a **negative** point charge (or a spherically symmetric charge)...
Gravitational potential, V_{grav}, is **potential energy** per **unit mass** and is **zero** at **infinity**.	Electric potential, $V_{electric}$, is **potential energy** per **unit positive charge** and is **zero** at **infinity**.

... and Three Differences too

Gravitational and electric fields aren't all the same — you need to know the **three main differences**:

1) Gravitational forces are always **attractive**. Electric forces can be either **attractive** or **repulsive**.
2) Objects can be **shielded** from **electric** fields, but not from gravitational fields.
3) The size of an **electric** force depends on the **medium** between the charges, e.g. plastic or air. For gravitational forces, this makes no difference.

Different materials have different permittivities. If you're not in free space (a vacuum), you'd use the permittivity of the substance, ε, instead of ε_0.

Practice Questions

Q1 What is meant by 'electric potential'?

Q2 State the formula for finding the electric potential in a radial field.

Q3 Sketch the equipotential surfaces around a point charge.

Q4 Write down the formula relating electric potential $V_{electric}$ and electric field strength $E_{electric}$.

Q5 Describe how $E_{electric}$ can be found from a graph of $V_{electric}$ against r, and how the change in $V_{electric}$ can be found from a graph of $E_{electric}$ against r.

Exam Questions

$e = -1.60 \times 10^{-19}$ C, $\varepsilon_0 = 8.85 \times 10^{-12}$ $C^2N^{-1}m^{-2}$ (Fm^{-1}).

Q1 Point A is 1.00 mm away from an electron. Calculate the electric potential at point A. [2 marks]

Q2 A spherical conductor with a radius of 3.5×10^{-3} m is charged such that its surface has an electric potential of 500.0 V. By modelling the sphere as a point charge, find its charge. [2 marks]

Q3 a) Two parallel plates are separated by a gap of 4.5 mm and connected to a 1500 V dc supply. What is the electric field strength between the plates? Give the unit and state the direction of the field. [2 marks]

b) The plates are then pulled further apart so that the distance between them is doubled. The voltage is adjusted so that the electric field strength remains the same. State the new voltage between the plates. [1 mark]

Q4 State two similarities and one difference between gravitational and electric fields. [3 marks]

I prefer gravitational fields — electric fields are repulsive...

Revising fields is a bit like a buy-one-get-one-free sale — you learn all about gravitational fields and they throw electric fields in for free. You just have to remember to change m for Q and G for $1/4\pi\varepsilon_0$... okay, so it's not quite a BOGOF sale. Maybe more like a buy-one-get-one-half-price sale... anyway, you get the point — go learn some stuff.

Millikan's Oil Drop Experiment

You'll probably know that the fundamental unit of charge, i.e. the size of the charge on an electron, is 1.60×10^{-19} C, and that all other charges are always exact multiples of this value. But at the beginning of the 20th century, many physicists thought that charge was a continuous variable that could take any value at all — Millikan showed this wasn't true.

Millikan's Experiment used *Stoke's Law*

1) Before you start thinking about Millikan's experiment, you need a bit of **extra theory**.
2) When you drop an object into a fluid, like air, it experiences a **viscous drag** force. This force acts in the **opposite direction** to the velocity of the object, and is due to the **viscosity** of the fluid.
3) You can calculate this viscous force on a spherical object using **Stoke's law**:

$$F = 6\pi\eta rv$$

where η is the viscosity of the fluid, r is the radius of the object and v is the velocity of the object.

Millikan's Experiment — the *Basic Set-Up*

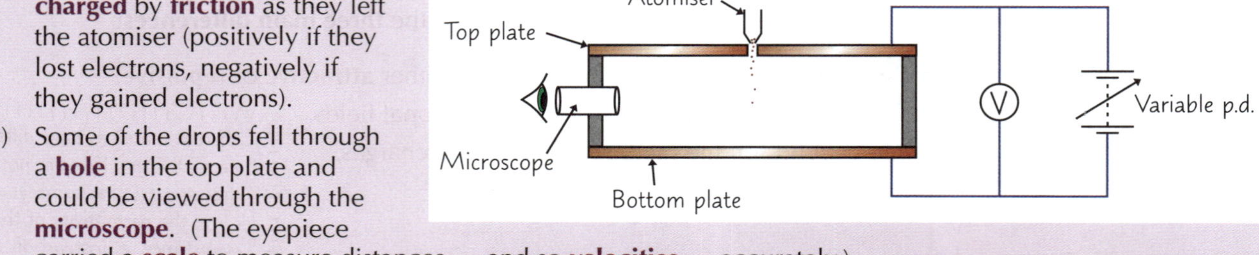

1) The **atomiser** created a **fine mist** of oil drops that were **charged** by **friction** as they left the atomiser (positively if they lost electrons, negatively if they gained electrons).
2) Some of the drops fell through a **hole** in the top plate and could be viewed through the **microscope**. (The eyepiece carried a **scale** to measure distances — and so **velocities** — accurately.)

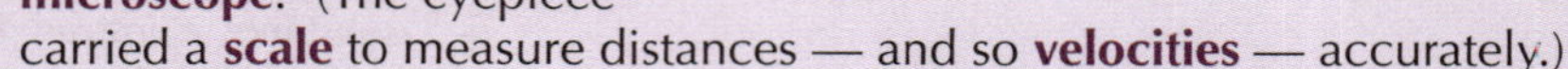

3) When he was ready, Millikan could apply a **potential difference** between the two plates, producing a **field** that exerted an upwards **force** on the charged drops. By **adjusting** the p.d., he could vary the strength of the field.

To give you a feel for the **size** of the apparatus, Millikan's plates were circular, with a diameter of about the width of this page. They were separated by about 1.5 cm.

Before the *Field* is Switched on, there's only *Gravity* and the *Viscous Force*

1) With the electric field turned off, the forces acting on each oil drop are:

   a) the **weight** of the drop (equal to mg) — acting downwards
   b) the **viscous force** from the air — acting upwards

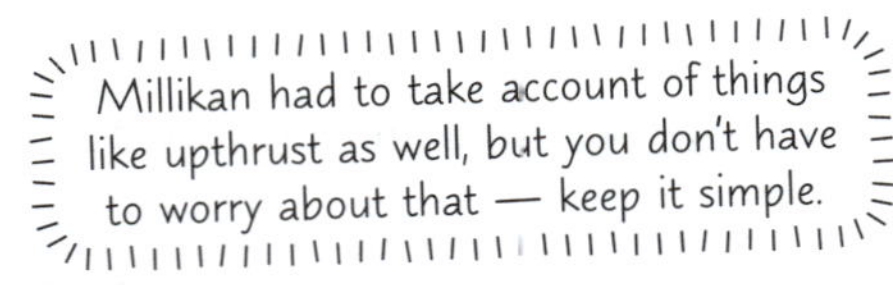

2) The drop will reach **terminal velocity** (i.e. it will stop accelerating) when these two forces are equal. So, from Stoke's law (see above):

$$mg = 6\pi\eta rv$$

It's assumed that the droplets are spherical for this experiment.

3) Since the **mass** of the drop is the **volume** of the drop multiplied by the **density**, ρ, of the oil, this can be rewritten as:

$$\frac{4}{3}\pi r^3 \rho g = 6\pi\eta rv \Rightarrow r^2 = \frac{9\eta v}{2\rho g}$$

Millikan measured η and ρ in separate experiments, so he could now calculate r — ready to be used when he switched on the electric field...

Millikan's Oil Drop Experiment

Then he **Turned On** the **Electric Field**...

1) The field introduced a **third major factor** — an **electric force** on each drop.
2) Millikan adjusted the applied p.d. until each drop was **stationary**. Since the **viscous force** is proportional to the **velocity** of the object, once each drop stopped moving, the viscous force **disappeared**.
3) Now the only two forces acting on each oil drop were:

   a) the **weight** of the drop — acting downwards
   b) the force due to the **uniform electric field** — acting upwards

Only the negatively charged droplets will experience an upwards force due to the electric field. The forces on the positively charged drops both act downwards, so the positively charged drops weren't of interest in this experiment.

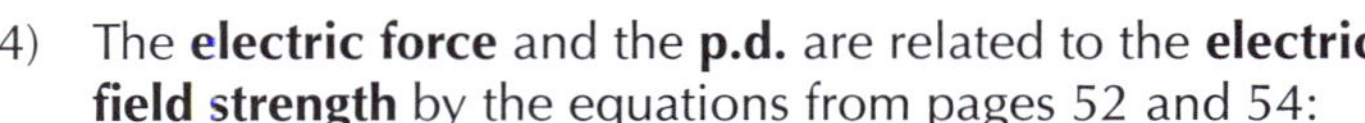

4) The **electric force** and the **p.d.** are related to the **electric field strength** by the equations from pages 52 and 54: $E_{electric} = \frac{F_{electric}}{q}$ $E_{electric} = \frac{V}{d}$
5) Combining these equations gives: $F_{electric} = \frac{qV}{d}$ where q is the charge on the oil drop, V is the p.d. between the plates and d is the distance between the plates.
6) Since the drop is **stationary**, this electric force must be equal to the weight, so: $\frac{qV}{d} = mg = \frac{4}{3}\pi r^3 \rho g$

   The first part of the experiment gave a value for r, so the **only unknown** in this equation is q.
7) So Millikan could use this equation to find the **charge on the drop**. He repeated the experiment for hundreds of drops — the charge on any drop was always a **whole number multiple** of $\mathbf{1.60 \times 10^{-19}}$ **C**.

These Results Suggested that **Charge was Quantised**

1) This result was **really significant**. Millikan concluded that charge can **never exist** in **smaller** quantities than 1.60×10^{-19} C. He assumed that this was the size of the **charge** carried by an **electron**.
2) Later experiments confirmed that **both** these things are true.

   Charge is "**quantised**".
   It exists in discrete "packets" of size **1.60×10^{-19} C** — the **fundamental unit of charge, e.**
   This is the size of the charge carried by **one electron**.

3) The charge on an electron is used to define a **unit of energy** — the **electron volt** (eV). 1eV = **1.60×10^{-19} J**.

Practice Questions

Q1 List the forces that act on an oil drop in Millikan's experiment:
   a) with the drop falling downwards at terminal velocity but with no applied electrical field,
   b) when the drop is stationary, with an electrical field applied.

Q2 Explain how Millikan's oil drop experiment provided evidence for the quantisation of the charge on an electron.

Exam Question

Q1 An oil drop of mass 1.63×10^{-14} kg is held stationary in the space between two charged plates 3.00 cm apart. The potential difference between the plates is 5000 V. The density of the oil used is 880 kg m^{-3}.

   a) Describe the relative magnitude and direction of the forces acting on the oil drop. [2 marks]
   b) Calculate the charge on the oil drop using $g = 9.81$ Nkg^{-1}. Give your answer in terms of e, the size of the charge on an electron. [3 marks]

   The electric field is switched off and the oil drop falls towards the bottom plate.

   c) Explain why the oil drop reaches terminal velocity as it falls. [3 marks]
   d) Calculate the terminal velocity of the oil drop using $\eta = 1.84 \times 10^{-5}$ kg m^{-1}s^{-1}. [3 marks]

So next time you need 1.5×10^{-19} coulombs — tough...

This was a huge leap. Along with the photoelectric effect this experiment marked the beginning of quantum physics.

Charged Particles in Magnetic Fields

Charged particles can be deflected by magnetic fields because the field exerts a force on the particles. This is the same effect that you saw in Module 6: Section 1 where a magnetic field exerted a force on a current-carrying wire.

Forces Act on Charged Particles in Magnetic Fields

Electric current in a wire is caused by the **flow** of negatively **charged** electrons. These charged particles are affected by **magnetic fields** — so a current-carrying wire experiences a **force** in a magnetic field (see page 46).

1) The equation for the **force** exerted on a **current-carrying wire** in a **magnetic field** perpendicular to the current is:

Equation 1: $F = BIl$

B is the magnetic flux density in teslas (T), *I* is the current in A, *l* is the length in m

2) To see how this relates to **charged particles** moving through a wire, you need to know that electric **current**, *I*, is the flow of **charge**, *q*, per unit **time**, *t*. $I = \frac{q}{t}$

3) A charged particle which moves a **distance *l*** in **time *t*** has a **velocity**, *v*:

$$v = \frac{l}{t} \Rightarrow t = \frac{l}{v}$$

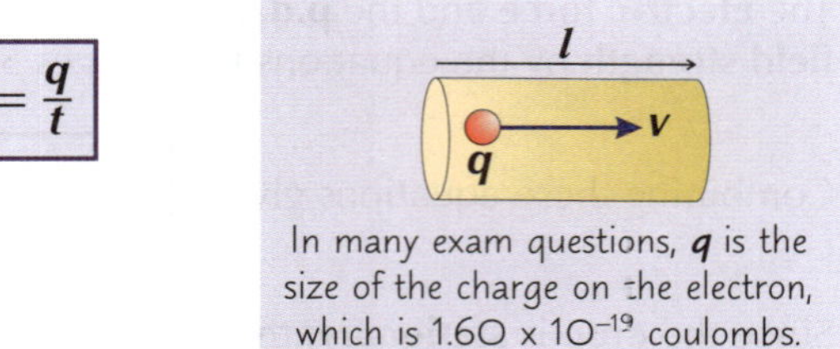

In many exam questions, *q* is the size of the charge on the electron, which is 1.60×10^{-19} coulombs.

4) Substituting this equation for time into the previous equation for current gives the **current** in terms of the **velocity** of the **charge** flowing through the **wire**:

Equation 2: $I = \frac{qv}{l}$

5) Putting **equation 2** back into **equation 1** gives the **electromagnetic force** on the wire as: $F = qvB$

6) You can use this equation to find the **force** acting on a **single charged particle moving through, and perpendicular to, a magnetic field**.

B is the magnetic flux density in teslas (T)
v is the particle velocity in ms^{-1}
q is the charge on the particle in C.

Example: An electron is travelling perpendicular to a uniform magnetic field of strength 2.0 T. If the electron is travelling at 2.0×10^4 ms^{-1} through the magnetic field, what is the force acting on the electron? (The charge on an electron is -1.60×10^{-19} C.)

$F = qvB$, so $F = (-1.60 \times 10^{-19}) \times (2.0 \times 10^4) \times 2.0 = -6.4 \times 10^{-15}$ N

Charged Particles in a Magnetic Field are Deflected in a Circular Path

1) By **Fleming's left-hand rule** (p.46) the force on a **moving charge** in a magnetic field is always **perpendicular** to its **direction of travel**.
2) Mathematically, that is the condition for **circular** motion (p.22).
3) This effect is found in **particle accelerators** such as **cyclotrons** and **synchrotrons** (see pages 68-69), which use **electric and magnetic fields** to accelerate particles to very **high energies** along circular paths.
4) The **radius of curvature** of the **path** of a charged particle moving through a magnetic field gives you information about the particle's **charge** and **mass** — this means you can **identify different particles** by studying how they're **deflected** (see next page).

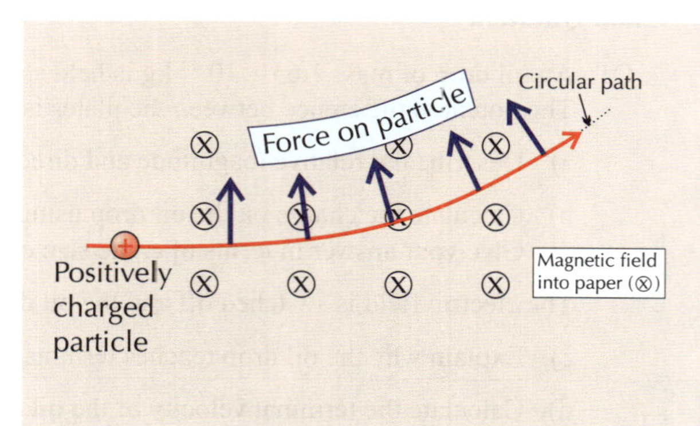

Charged Particles in Magnetic Fields

Centripetal Force Tells Us About a Particle's Path

The centripetal force and the electromagnetic force are equivalent for a charged particle travelling along a circular path.

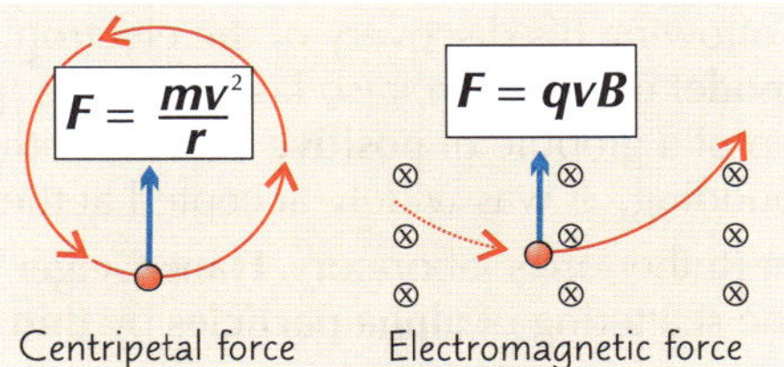

1) For uniform circular motion **Newton's second law** gives:

$$F = \frac{mv^2}{r}$$

2) So, for a **charged particle** following a **circular** path in a **magnetic field** (where $F = qvB$):

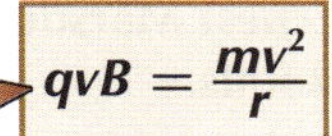

$$qvB = \frac{mv^2}{r}$$

3) Rearranging gives:

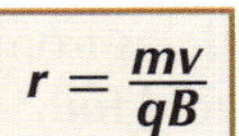

$$r = \frac{mv}{qB}$$

Where: m is the mass of the particle, v is its speed and r is the radius of the circular path.

4) So, different charged particles will have paths with different radii — the higher the ratio of m to q, the larger the radius of the path.

Eric and Phil were *not* lost. They'd just accidentally demonstrated the path of a charge in a magnetic field.

Example: A magnetic field of strength 0.080 T is used to move an electron in a circular path with a radius of 1.8×10^{-4} m at a constant speed. Calculate the speed of the electron.

1) The force on a charged particle moving in a magnetic field, $F = qvB$.
2) The particles move in a circle, so F gives the centripetal force $\Rightarrow qvB = \frac{mv^2}{r}$
3) Rearranging for v gives: $v = \frac{qBr}{m}$
4) Substitute the values: $v = \frac{(1.60 \times 10^{-19}) \times 0.080 \times (1.8 \times 10^{-4})}{9.11 \times 10^{-31}} = 2.5 \times 10^6$ ms^{-1} (to 2 s.f.)

The size of the charge on an electron is 1.60×10^{-19} C (only the magnitude of the charge is used here — we're calculating speed, not velocity).

The mass of an electron is 9.11×10^{-31} kg.

The charge and mass of an electron are given in the data and formulae booklet in the exam.

Practice Questions

Q1 Write down an equation to calculate the force on a charged particle moving in a magnetic field.

Q2 Explain why the force experienced by a moving charge in a magnetic field causes it to move with circular motion.

Q3 Explain how charged particles can be identified as they move through a magnetic field.

Exam Questions

$m_e = 9.11 \times 10^{-31}$ kg, e = magnitude of the charge on an electron = 1.60×10^{-19} C

Q1 a) An electron travels at a velocity of 5.00×10^6 ms^{-1} through a perpendicular magnetic field of 0.770 T. Find the magnitude of the force acting on the electron. [2 marks]

b) An electric field is then applied, which causes the electron to travel in a straight line through the magnetic field. Calculate the electric field strength of the electric field. [2 marks]

Q2 An electron is accelerated to a velocity of 2.3×10^7 ms^{-1} by a particle accelerator. The electron moves in a circular path perpendicular to a magnetic field of 0.6 mT.

a) Use the equations for electromagnetic and centripetal force to show that $qB = \frac{mv}{r}$. [2 marks]

b) Use the relation from part a) to calculate the radius of the electron's path. [1 mark]

Hold on to your hats folks — this is starting to get tricky...

Basically, the main thing you need to know here is that a magnetic field will exert a force on a charged particle, making it follow a circular path. There's even a handy equation to work out the force on a charged particle moving through a magnetic field — it might not impress your friends, but it will impress the examiner, so learn it.

Scattering to Determine Structure

By firing radiation at different materials, you can take a sneaky beaky at their internal structures...

Rutherford's Experiment Disproved the Thomson Model

1) Following his discovery of the electron in the late 19th century, **J.J. Thomson** proposed the **Thomson model** of the atom, also known as the '**plum pudding**' model. This model said that atoms were made up of a globule of **positive charge**, with **negatively charged electrons sprinkled** in it, like fruit in a plum pudding. It was widely accepted at the time, until the **Rutherford scattering experiment** of 1909.
2) In Rutherford's laboratory, **Hans Geiger** and **Ernest Marsden** studied the scattering of **alpha particles** by **thin metal foils**.

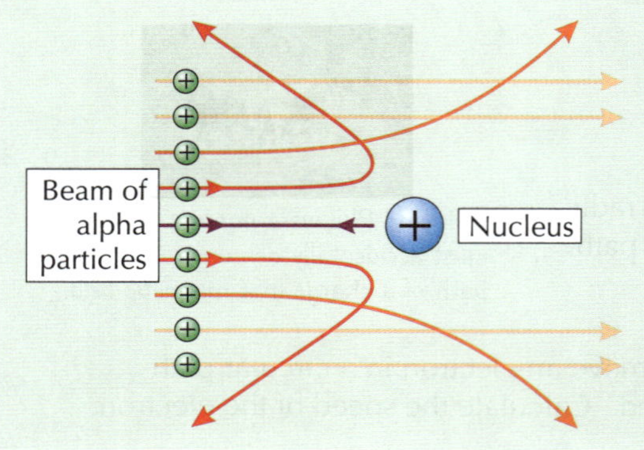

1) A **stream of alpha particles** from a radioactive source was fired at **very thin gold foil**.
2) Geiger and Marsden recorded the **number** of alpha particles scattered at **different angles**.
3) Geiger and Marsden observed that alpha particles occasionally **scatter at angles greater than 90°**. This can only be possible if they're **striking something more massive** than themselves.

Rutherford's Model of the Atom — The Nuclear Model

Alpha particles are made up of two protons and two neutrons, so they have an overall positive charge (see p.72).

This experiment led Rutherford to some **important conclusions**:

1) Most of the fast, charged alpha particles went **straight through** the foil. So the atom is mainly **empty space**.
2) **Some** of the alpha particles were **deflected** through **large angles**, so the **centre** of the atom must have a **large**, **positive charge** to repel them. Rutherford named this the **nucleus**.
3) Very few particles were deflected by angles greater than **90 degrees**, so the nucleus must be **tiny**.
4) Most of the **mass** must be in the nucleus, since the fast alpha particles (with high momentum) are deflected by the nucleus.

So most of the **mass** and the **positive charge** in an atom must be contained within a **tiny**, **central** nucleus.

When an alpha particle gets close enough to the nucleus, the electrostatic force (see p.52) between the positively charged nucleus and the positively charged alpha particle results in a repulsion between the two. This is what causes some of the alpha particles to scatter.

Atoms are made up of Protons, Neutrons and Electrons

Inside **every atom**, there's a **nucleus** containing **protons** and **neutrons**.
Protons and **neutrons** are both known as **nucleons**. **Orbiting** this core are the **electrons**.
This is the **nuclear model** of the atom.

The diagram shows **neutral oxygen**, with **eight protons** and **eight electrons**.

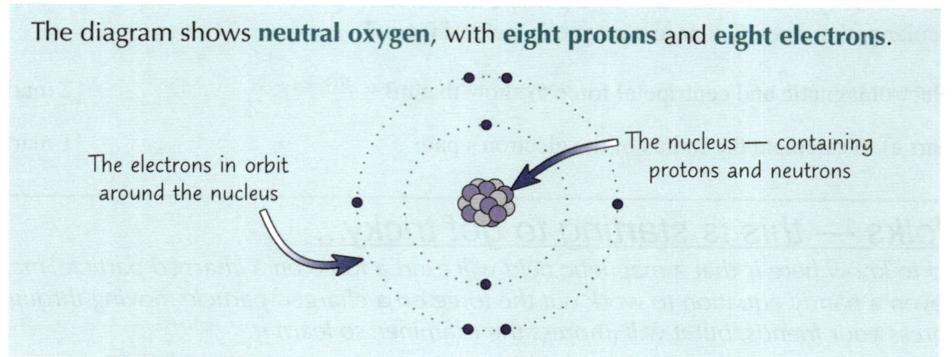

Forget Rutherford — James reckoned his model would be the best yet.

Scattering to Determine Structure

You can Estimate the Closest Approach of a Scattered Particle

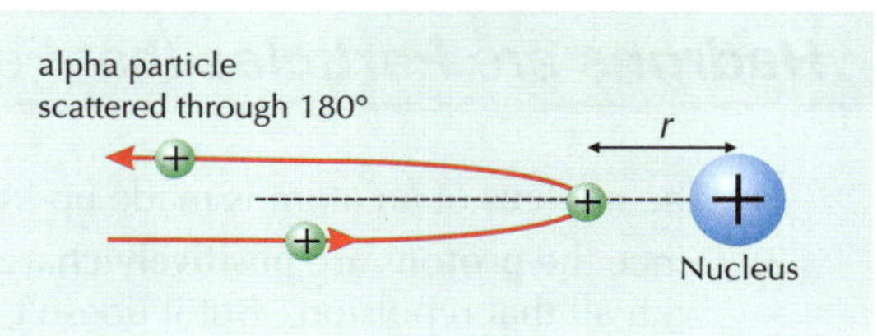

1) When you fire an alpha particle at a gold nucleus, you can estimate its distance of **closest approach** if you know the alpha particle's **initial kinetic energy**.
2) An alpha particle that 'bounces back' and is deflected through 180° will have momentarily stopped a short distance from the nucleus. It does this at the point where its **electric potential energy** (see p.53) **equals** its **initial kinetic energy**.
3) It's just conservation of energy — and you can use it to find how close the particle can get to the nucleus:

$$\textbf{Initial K.E. = electric potential energy} = \frac{kq_{alpha}Q_{gold}}{r} \quad \text{where } k = \frac{1}{4\pi\varepsilon_0}$$

You might also see k given as 8.98×10^9 Nm^{2}C^{-2} (to 3 s.f.)

Q_{gold} is the charge on a gold nucleus in C, q_{alpha} is the charge on an alpha particle in C, ε_0 is the permittivity of free space, 8.85×10^{-12} Fm^{-1}

4) The **distance of closest approach** is given by r in the equation above.
5) To find the charge of a nucleus, you need to know the atom's **proton number**, ***Z*** — that tells you how many protons are in the nucleus.
6) A proton has a charge of **+e** (where e is the size of the charge on an electron, 1.60×10^{-19} C), so the charge of a nucleus must be **+*Z*e**. The charge on an alpha particle is **+2e**.

If you need to calculate the force experienced by the particle as it approaches the nucleus, remember $F_{electric} = \frac{kqQ}{r^2}$ (p.52).

Example: An alpha particle with an initial kinetic energy of 6.0 MeV is fired at a gold nucleus ($Z_{gold} = 79$). Calculate the closest approach of the alpha particle to the nucleus.

Initial K.E. = 6.0 MeV = 6.0×10^6 eV

Convert this energy into joules: $6.0 \times 10^6 \times 1.60 \times 10^{-19} = 9.6 \times 10^{-13}$ J

$1 \text{ eV} = 1.60 \times 10^{-19}$ J.

So, at the distance of closest approach, electric potential energy $= \frac{kq_{alpha}Q_{gold}}{r} = \frac{q_{alpha}Q_{gold}}{4\pi\varepsilon_0 r} = 9.6 \times 10^{-13}$ J

Rearrange to get $r = \frac{q_{alpha}Q_{gold}}{9.6 \times 10^{-13} \times 4\pi\varepsilon_0} = \frac{(2 \times 1.60 \times 10^{-19}) \times (79 \times 1.60 \times 10^{-19})}{(9.6 \times 10^{-13}) \times 4\pi \times (8.85 \times 10^{-12})} = 3.788... \times 10^{-14}$

$= \mathbf{3.8 \times 10^{-14}}$ **m (to 2 s.f.)**

Practice Questions

Q1 Sketch a diagram to show the paths of alpha particles when they are scattered by a nucleus in Rutherford's gold foil experiment.

Q2 Describe the structure of the atom according to the nuclear model.

Q3 Explain how alpha-particle scattering gives evidence for a small massive nucleus.

Exam Questions

Q1 A beam of alpha particles is directed onto a very thin gold film.

a) Explain how alpha particles are scattered by atomic nuclei. [3 marks]

b) Explain why the majority of the alpha particles are not scattered. [1 mark]

Q2 A proton is fired at an aluminium nucleus ($Z_{aluminium} = 13$). It has an initial kinetic energy of 4.0 MeV. Show that the closest approach of the proton to the nucleus is 4.7×10^{-15} m (to 2 s.f.). [4 marks]

Alpha scattering — it's positively repulsive...

The important things to learn from these two pages are the nuclear model for the structure of the atom (i.e. a large mass nucleus surrounded by orbiting electrons) and how Geiger and Marsden's alpha-particle scattering experiment gives evidence that supports this model. Once you know that, take a deep breath — it's about to get a little more confusing.

Particles and Antiparticles

There are loads of different types of particle apart from the ones you get in normal matter (protons, neutrons, etc.). They only appear in cosmic rays and in particle accelerators, and they often decay very quickly.

***Hadrons** are **Particles** that Feel the **Strong Interaction** (e.g. Protons and Neutrons)*

1) The **nucleus** of an atom is made up from **protons** and **neutrons** (this is sounding familiar...).
2) Since the **protons** are **positively charged** you might think that the nucleus would **fly apart** with all that repulsion. But it doesn't, so there has to be a strong **force** holding the **p**'s and **n**'s together.
3) That force is called the **strong interaction** (who said physicists lack imagination...).
4) Not all particles can feel the strong interaction — the ones that can are called **hadrons**.
5) Hadrons aren't **fundamental** particles. They're made up of **smaller particles** called **quarks** (see page 66).
6) As well as **protons** and **neutrons**, there are **other hadrons** that you don't get in normal matter, like **sigmas** (Σ) and **mesons** — luckily you **don't** need to know about them (woohoo!).

*The **Proton** is the **Only Stable Hadron***

1) Most **hadrons** will eventually **decay** into **other particles**.
   The exception is protons — most physicists think that protons don't **decay**.
2) The **neutron** is an **unstable particle** that **decays** into a **proton**.
   (But it's much more stable when it's part of a nucleus.) It's really just an **example** of **β⁻ decay** (see p.74), which is caused by the **weak interaction**.
   The particle reaction for this decay is:

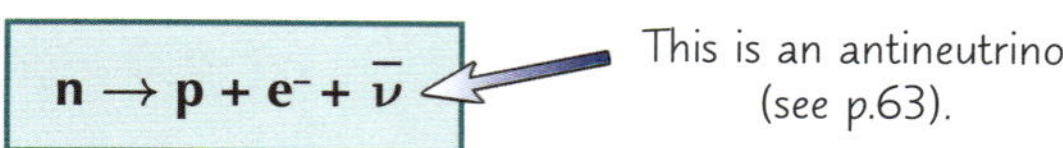

$$n \rightarrow p + e^- + \bar{\nu}$$

3) Free neutrons (i.e. ones not held in a nucleus) have a half-life of about **15 minutes**.

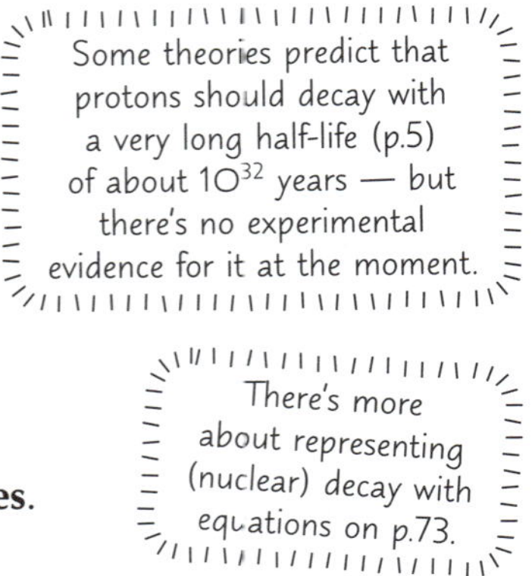

***Leptons Don't** Feel the **Strong Interaction** (e.g. Electrons and Neutrinos)*

1) **Leptons** are **fundamental particles** that **don't** feel the **strong interaction**. They **interact** with other particles via the **weak interaction** and **gravity** (and the electromagnetic force if they're charged).
2) There are two types of lepton you need to know about — **electrons (e⁻)**, which should be familiar to you, and **neutrinos (ν)**.
3) Neutrinos have **zero** (or almost zero) **mass** and **zero electric charge** — so they don't do much. **Neutrinos** only take part in **weak interactions** (see p.67). In fact, a neutrino can **pass right through the Earth** without **anything** happening to it.
4) **Lepton number** is the number of leptons.
   Electrons and neutrinos have a lepton number of **+1**.
5) The total lepton number in any particle reaction **never changes** — **lepton number is conserved**. For example, in the neutron decay reaction above, n and p have lepton number 0, e⁻ has lepton number 1, and $\bar{\nu}$ has lepton number –1 (see next page).
   So 0 → 0 + 1 – 1. There's more on the **properties** of **different particles** on the next page.

Name	Symbol	Charge (relative to the fundamental charge, e)
electron	e⁻	–1
neutrino	ν	0

ν is the Greek letter "nu".

Antiparticles** were **Predicted** Before they were **Discovered

When **Paul Dirac** wrote down an equation obeyed by **electrons**, he found a kind of **mirror image** solution.

1) It predicted the existence of a particle like the **electron** but with **opposite electric charge** — the **positron**.
2) The **positron** turned up later in a cosmic ray experiment.
   Positrons have **identical mass** to electrons but they carry a **positive** charge.

Particles and Antiparticles

Every Particle has an Antiparticle

Each particle type has a **corresponding antiparticle** with the **same mass** but with **opposite charge**.
For instance, an **antiproton** is a **negatively charged** particle with the same mass as the **proton**.

Even the shadowy **neutrino** has an antiparticle version called the **antineutrino** — it doesn't do much either.

Particle	Symbol	Relative charge	Rest mass / kg	Lepton no.	Antiparticle	Symbol	Relative charge	Rest mass / kg	Lepton no.
proton	p	+1	1.673×10^{-27}	0	antiproton	$\bar{p}$	–1	1.673×10^{-27}	0
neutron	n	0	1.675×10^{-27}	0	antineutron	$\bar{n}$	0	1.675×10^{-27}	0
electron	e^-	–1	9.11×10^{-31}	1	positron	e^+	+1	9.11×10^{-31}	–1
neutrino	ν	0	0	1	antineutrino	$\bar{\nu}$	0	0	–1

1) In the exam, you'll be **given** the masses of protons, neutrons and electrons. Just remember that the mass of an **antiparticle** is the **same** as the mass of its corresponding particle.
2) The masses in the table are all **rest masses** — the mass of the particle when it's **not moving**. This is because the masses of objects change when they're moving at very high speeds, but you don't need to know about that.
3) You need to **learn** the **relative charge** on each type of particle (these are all relative to $e = 1.60 \times 10^{-19}$ C).
4) In any particle reaction **charge is conserved** — the total charge before the reaction must be equal to the total charge after the reaction.
5) Neutrinos are **incredibly tiny** — you can assume they have zero mass and zero charge.
6) **Anti-leptons**, like the positron and antineutrino, have a **lepton number of –1**. So whenever a lepton is produced in a particle reaction, an anti-lepton is produced too (to conserve lepton number).

Practice Questions

Q1 Give two differences between a hadron and a lepton.

Q2 What is the only stable hadron? Name another hadron that will decay into it.

Q3 Name two types of lepton. Give their lepton numbers.

Q4 Which antiparticle has zero charge and a rest mass of 1.675×10^{-27} kg?

Q5 What is the symbol for an antineutrino? List the properties of an antineutrino.

Q6 Write down the charge of a positron, given that the charge on an electron is -1.60×10^{-19} C.

Q7 Give one similarity and one difference between a proton and an antiproton.

Exam Questions

Q1 State the decay products of the neutron.
Explain why this decay cannot be due to the strong interaction. [3 marks]

Q2 A particle is detected that has no charge and that does not feel the strong interaction.
Suggest what the particle might be. [1 mark]

Q3 a) Explain why the reaction $\bar{n} \rightarrow \bar{p} + e^+ + \bar{\nu}$ is not possible. [2 marks]

b) Explain why the reaction $\bar{n} \rightarrow \bar{p} + e^- + \bar{\nu}$ is not possible. [2 marks]

Go back to the top of page 62 — do not pass GO, do not collect £200...

There's a frankly silly number of physics words on this page, but it looks worse than it is. Honestly. Give it another read, and don't move on until you're sure you know it all, otherwise the next few pages will sound like nonsense...

Pair Production and Annihilation

Time for one of physics' most famous equations. I know you're on the edge of your seat. I certainly I am...

You can Create Matter and Antimatter from Energy

You've probably heard about the **equivalence** of energy and mass. It all comes out of Einstein's special theory of relativity. **Energy** can turn into **mass** and **mass** can turn into **energy** if you know how — all you need is one fantastic and rather famous formula:

$$E_{rest} = mc^2$$

E_{rest} is the 'rest energy' — the energy equivalent to the mass, m, when it isn't moving. Energy is also approximately equal to mc^2 for masses moving at slow speeds, but things get a bit more complicated when objects are moving near the speed of light (p.69).

As you've probably guessed, there's a bit **more to it** than that:

When **energy** is converted into **mass** you get **equal amounts** of **matter** and **antimatter**.

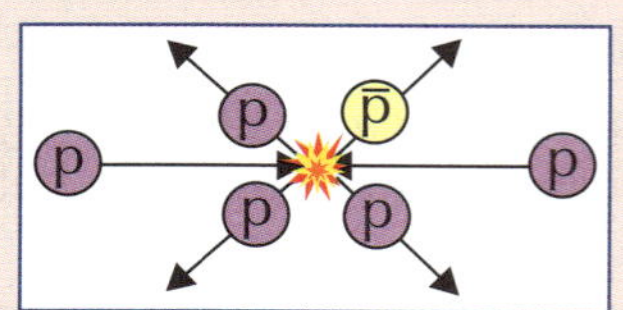

Fire **two protons** at each other at high speed and you'll end up with a lot of **energy** at the point of impact. This energy might be converted into **more particles**.

If an extra **proton** is formed then there will always be an **antiproton** to go with it. It's called **pair production**.

Each Particle-Antiparticle Pair is Produced from a Single Photon

Pair production only happens if **one gamma ray photon** has enough energy to produce that much mass. It also tends to happen near a **nucleus**, which helps conserve momentum.

You usually get **electron-positron** pairs produced (rather than any other pair) — because they have a relatively **low rest mass**.

You can calculate the **minimum energy** (and therefore the minimum frequency and maximum wavelength) a photon must have for pair production to occur using $E_{rest} = mc^2$.

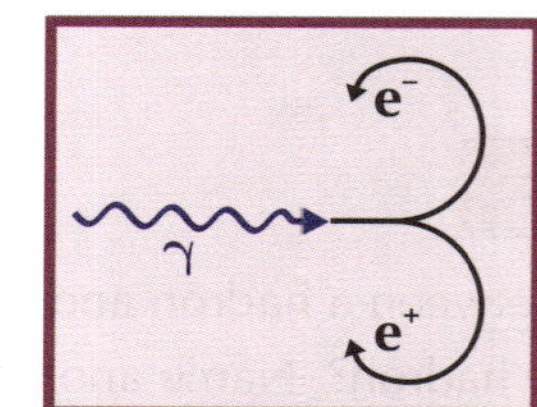

The particle tracks are curved because there's usually a magnetic field present in particle physics experiments (see next page).

Example: An electron and a positron are produced from a single photon. Find the maximum possible wavelength of the photon. (The rest mass of an electron $m_e = 9.11 \times 10^{-31}$ kg.)

1) As **energy before = energy after**, the **minimum energy** the photon can have is equal to the energy needed to produce the particles at rest (i.e. the particles have no kinetic energy).
2) An electron and a positron have the **same rest mass** (m_e), which means that:
$E_{photon} = E_{rest(electron)} + E_{rest(positron)} = 2m_ec^2$
This is the minimum energy the photon needs to create the two particles. If the photon has more energy than this, then it could be converted into kinetic energy of the electron and positron, or used to create a more massive particle/antiparticle pair instead.
3) The **energy** of a photon is related to its **wavelength** by the equation: $E_{photon} = \frac{hc}{\lambda}$ ← h is the Planck constant (6.63×10^{-34} Js).
4) Just put these two equations for E_{photon} together and **rearrange** to find λ:

$$2m_ec^2 = \frac{hc}{\lambda} \text{ so } \lambda = \frac{hc}{2m_ec^2} = \frac{h}{2m_ec}, \text{ and so: } \lambda = \frac{6.63\times10^{-34}}{2\times(9.11\times10^{-31})\times(3.00\times10^{8})}$$

$$= 1.212... \times 10^{-12} = \mathbf{1.21 \times 10^{-12}\ m\ (to\ 3\ s.f.)}$$

This is in the gamma part of the EM spectrum — gamma radiation has a wavelength of approximately 10^{-12} m or less.

Pair Production and Annihilation

The Opposite of Pair Production is Annihilation

1) When a **particle** meets its **antiparticle** the result is **annihilation**.
2) All the **mass** of the particle and antiparticle gets converted to **energy**, in the form of a pair of identical photons.
3) Antiparticles can generally only exist for a **fraction of a second** before they annihilate, so you won't see many of them.
4) Just like with pair production, you can calculate the **minimum energy** of each photon produced (i.e. assuming that the particles have **negligible** kinetic energy).
5) The combined energy of the photons will be equal to the combined energy of the particles, so $2E_{photon} = 2mc^2$ and so:

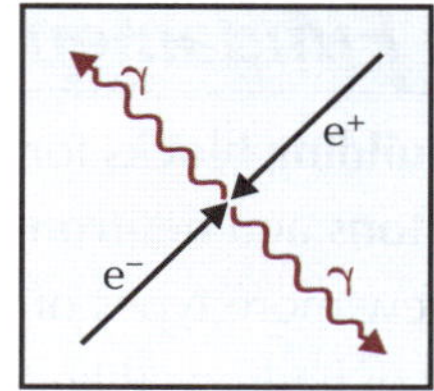

The electron and positron annihilate and their mass is converted into the energy of a pair of identical gamma ray photons.

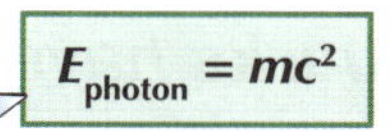

You can calculate the minimum frequency and maximum wavelength as before (i.e. by using $E = hf = \frac{hc}{\lambda}$).

You Can Use a Cloud Chamber to Observe Pair Production and Annihilation

1) A cloud chamber is basically a large box filled with alcohol (or water) **vapour**.
2) When high energy particles pass through the cloud chamber, they **ionise** alcohol particles along their path.
3) The vapour in the cloud chamber **condenses** around these ions, forming a **trail of alcohol droplets** ('a cloud') along the path of the charged particle.
4) Most cloud chambers include a **magnetic field** at **right-angles** to the direction of particle motion. A moving charge in a magnetic field experiences a force (p.58), so the paths of **charged particles**, like electrons and positrons, will **bend** as they pass through the chamber. Positively and negatively charged particles are deflected in **opposite directions**.
5) As particles travel through a cloud chamber, they **slow down**, as ionising alcohol particles uses energy. This means the paths of charged particles when a magnetic field is present are **spirals**, rather than circles.
6) Gamma ray photons are only **weakly ionising** (p.72) — they pass through the alcohol of the cloud chamber without interacting with it very much. This means they don't produce trails.

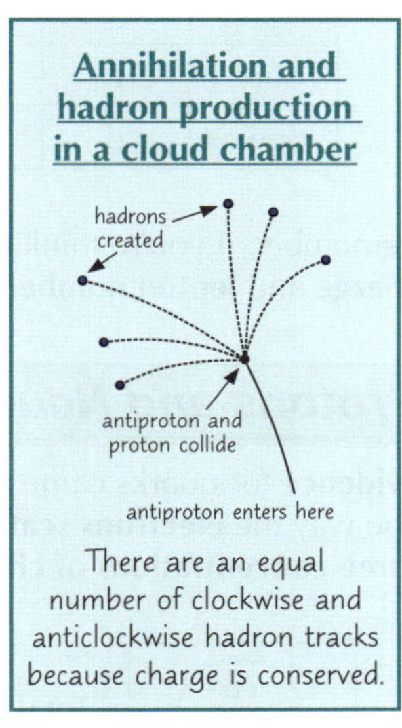

There are an equal number of clockwise and anticlockwise hadron tracks because charge is conserved.

Practice Questions

Q1 Write down Einstein's famous equation. Define each term used.

Q2 A high energy photon produces a positron. What other particle must be produced?

Q3 Briefly explain how a cloud chamber works.

Q4 What effect does the presence of the vapour in a cloud chamber have on the path of a charged particle if the particle's path is bending due to an applied magnetic field?

Exam Questions

Q1 Explain why the reaction $p + p \rightarrow p + p + n$ is not possible. [1 mark]

Q2 Write down an equation for the reaction between a positron and an electron and state the name of this type of reaction. [2 marks]

Q3 Assuming both particles have negligible kinetic energy, calculate the frequency of the two photons produced when a proton and an antiproton annihilate.
($m_p = 1.673 \times 10^{-27}$ kg, $h = 6.63 \times 10^{-34}$ Js, $c = 3.00 \times 10^8$ ms^{-1}) [3 marks]

Pair production — never seems to happen with my socks...

Learn the key points on these pages and it'll all be plain sailing. And don't forget: a) if energy is converted into a particle, you also get an antiparticle, b) an antiparticle won't last long before it bumps into the right particle and annihilates with it, c) annihilation releases the energy it took to make them to start with...

Quarks

Quarks are the fundamental particles that make up protons and neutrons. If you haven't yet, it's probably best to read p.62–63 before you start — then this will all make a bit more sense...

Quarks *are* ***Fundamental Particles***

Quarks are the **building blocks** for **hadrons** like **protons** and **neutrons**.

1) To make **protons** and **neutrons** you only need two types of quark — the **up** quark (**u**) and the **down** quark (**d**).
2) There are a few more types of quark, but you don't need to know about them (woohoo!).

The **antiparticles** of hadrons (like antiprotons and antineutrons) are made from **anti-quarks**.

Quarks *and* ***Anti-quarks*** *have* ***Opposite Charge***

The **anti-quarks** have **opposite charges** to the quarks — as you'd expect.

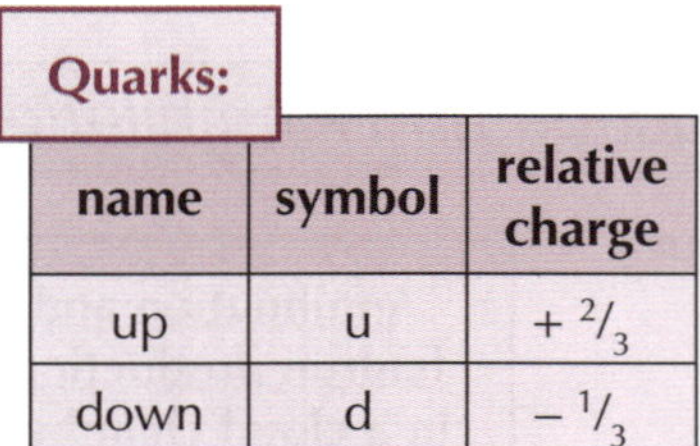

Quarks:

name	symbol	relative charge
up	u	$+\frac{2}{3}$
down	d	$-\frac{1}{3}$

Anti-quarks:

name	symbol	relative charge
anti-up	$\bar{u}$	$-\frac{2}{3}$
anti-down	$\bar{d}$	$+\frac{1}{3}$

As on p.63, these charges are relative to the elementary charge, *e*.

Remember, if you're thinking about particle reactions in terms of quarks, **charge** and **lepton number** still need to be **conserved**.

Protons and Neutrons *are Made from* ***Three Quarks***

Evidence for quarks came from **hitting protons** with **high energy electrons**. The way the **electrons scattered** showed that there were **three concentrations of charge** (quarks) **inside** the proton.

The quarks that a particle is made up from is called its 'quark composition'.

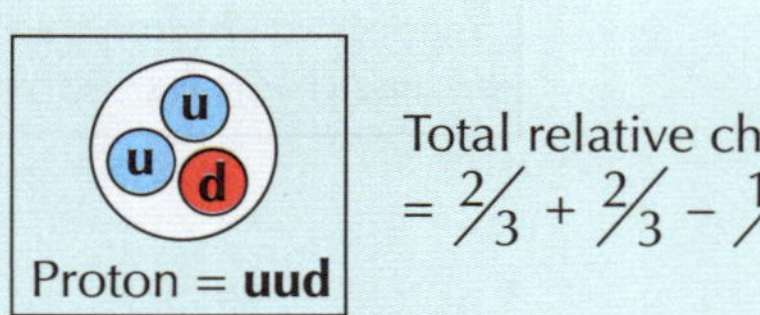

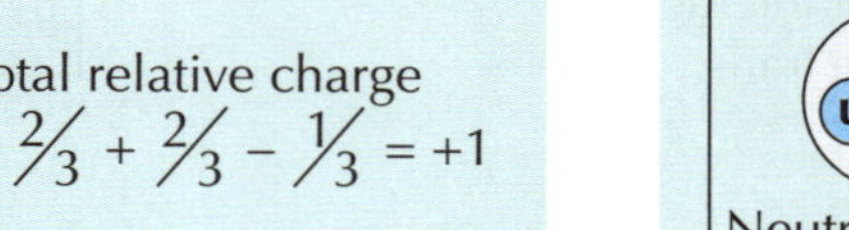

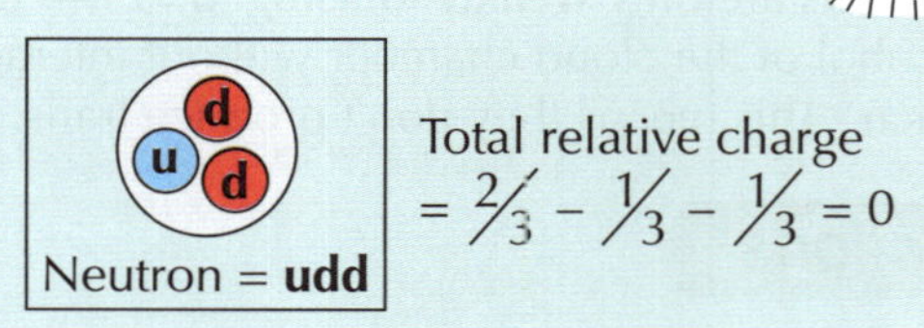

Antiprotons are $\bar{u}\bar{u}\bar{d}$ and antineutrons are $\bar{u}\bar{d}\bar{d}$ — so no surprises there then.

Not all hadrons have **three quarks** though. Protons and neutrons are a type of hadron called a **baryon**, which is made up of **three quarks**. There are also hadrons made up of a **quark** and an **anti-quark**, called **mesons** — but you don't really need to know about them.

There's ***no Such Thing*** *as a* ***Free Quark***

What if you **blasted** a **proton** with **enough energy** — could you **separate out** the quarks? Nope. The energy just gets changed into more **quarks and antiquarks** — it's **pair production** again (see p.64) and it makes **mesons**.

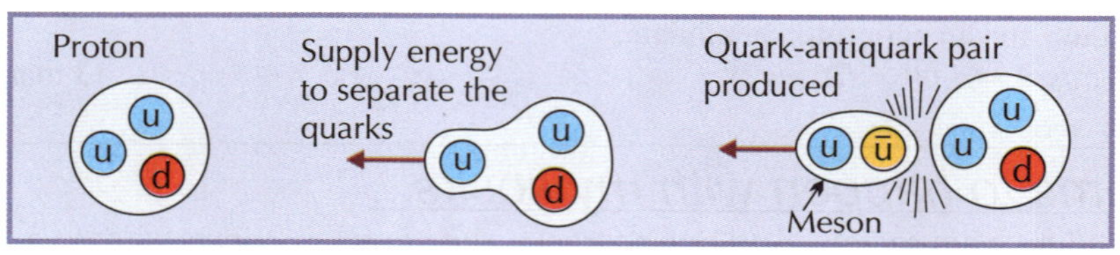

This is called **quark confinement**.

Quarks

Gluons Provide Force Between Quarks

1) When two particles **interact**, something must **happen** to let one particle know that the other one's there. That's the idea behind **exchange particles**. You can picture them if you think about **balls** and **boomerangs**:

Repulsion — Each time the **ball** is **thrown or caught** the people get **pushed apart**. It happens because the ball carries **momentum**.

Attraction — Each time the **boomerang** is **thrown or caught** the people **get pushed together**. (In real life, you'd probably fall in first.)

The particles don't actually loop round like that, though.

These exchange particles are called **gauge bosons** — they're virtual particles that only last for a very short time.

2) **All forces in nature** are caused by four **fundamental** forces. Each one has its **own gauge boson**:

Type of Interaction	Gauge Boson	Particles Affected
strong	gluon	hadrons only
electromagnetic	photon	charged particles only
weak	W^+, W^-, Z^0	all types
gravity	graviton?	all types

Particle physicists never bother about gravity because it's so incredibly feeble compared with the other types of interaction. Gravity only really matters when you've got big masses like stars and planets. The graviton may exist but there's no evidence for it.

3) The exchange particle that causes the **strong force** that 'glues' hadrons like protons and neutrons together is imaginatively called the **gluon**.
4) Because gluons cause a force, you can think of them as **fields** as well as particles. It's just the same as thinking of a **gravitational force** as being caused by a **gravitational field**.
5) As you try to **separate** quarks, you actually **increase** the **energy** of the gluon field, **increasing** the **attraction** between them.

quark — gluon field — quark

6) If you keep pulling, eventually the energy in the gluon field will be enough that it produces a **quark-antiquark pair**. This is why you can **never** detect a quark on its own (see p.66).

Gluon field energy increases as you separate the quarks. → Eventually a quark-antiquark pair is produced. + quark antiquark

Practice Questions

Q1 What is a quark?

Q2 What is the relative charge on an anti-down quark?

Q3 Explain why quarks are never observed on their own.

Q4 Name the exchange particle for the strong force felt between two quarks.

Exam Questions

Q1 State the combination of three quarks that make up a neutron. [1 mark]

Q2 Give the quark composition of the proton. Explain how the relative charge of each quark gives rise to its total relative charge. [2 marks]

A quark — not the noise a posh duck makes...

Don't know about you, but I'm getting a wee bit sick of tables of particles to learn. Sadly you need to know all this stuff, so it's time to make yourself a cuppa and give it all another read...

Particle Accelerators

Particle accelerators are devices that (surprisingly) accelerate particles, using electric and magnetic fields. Accelerated particles can be used in scattering experiments to investigate the fundamental particles that make up matter...

Particle Accelerators** Cause High-Energy **Collisions

There are lots of different types of accelerator out there smashing particles together. One of the main types is the linear accelerator...

A linear particle accelerator

negatively charged particle

electrodes

alternating p.d. changes the charge on each electrode

1) A **linear accelerator** is a long **straight** tube containing a series of **electrodes**. The charge on each electrode alternates along the tube (so a positive electrode is always between two negative electrodes and vice-versa).
2) The electrodes are connected to an **alternating p.d. supply** so that the charge of each **electrode** continuously **changes** between + and –.
3) The alternating p.d. is **timed** so that the particles are always **attracted** to the **next electrode** in the accelerator and **repelled** from the **previous** one.
4) A particle's **speed increases** each time it **passes** an electrode — so if the accelerator is long enough particles can be made to approach the **speed of light**.
5) The **high-energy particles** leaving a linear accelerator **collide** with a **fixed target** at the end of the tube.

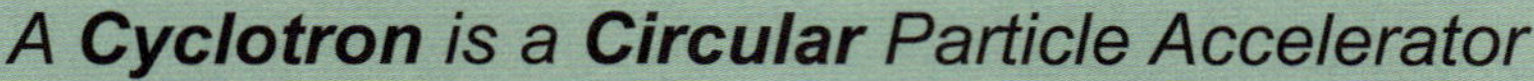

*A **Cyclotron** is a **Circular** Particle Accelerator*

1) A cyclotron uses **two semicircular electrodes** to accelerate protons or other charged particles across a gap.
2) An **alternating potential difference** is applied between the electrodes — as the **particles** are **attracted** from one side to the other their **energy increases** (i.e. they are **accelerated**).
3) A **magnetic field** is used to keep the particles moving in a **circular motion** (in the diagram on the right, the magnetic field would be perpendicular to the page).
4) The combination of the **electric** and **magnetic fields** makes the particles **spiral outwards** as their energy increases.

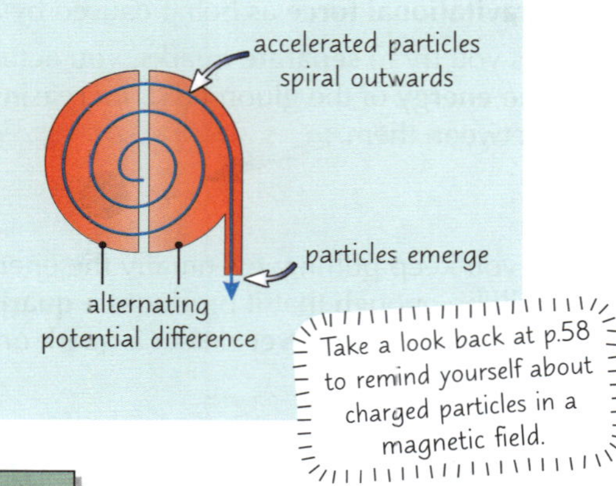

Take a look back at p.58 to remind yourself about charged particles in a magnetic field.

***Synchrotrons** Produce **Very High Energy** Beams*

1) A **synchrotron** can produce particle collisions with much **higher energies** than either a linear accelerator or a cyclotron.
2) **Electromagnets** keep the particles moving in a **circular path** in **focused beams**.
3) In this way, **synchrotrons** can produce particles with energies reaching from **500 GeV to several TeV**.
4) You can find the **force** experienced by a particle in a synchrotron (or a cyclotron) due to the **magnetic field** using the formula below (see p.58).

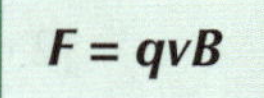

$$F = qvB$$

F is the force on the particle (in N), q is the charge on the particle (in C), v is its velocity in (ms^{-1}), and B is the magnetic flux density (in T).

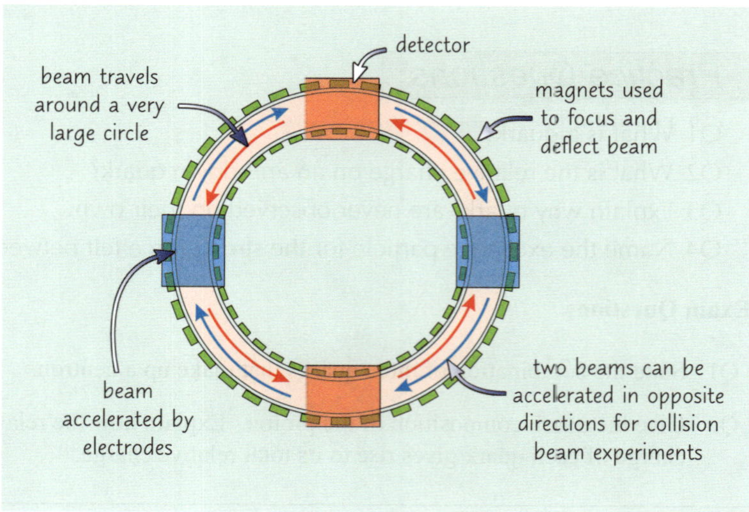

5) Synchrotrons like the Large Hadron Collider (LHC) at CERN in Switzerland are **really expensive** to build and run, so they need to be funded **internationally**.

Particle Accelerators

Masses **Cannot** Reach the Speed of Light

1) Particles can be accelerated to such high speeds that the effects of **relativity** become noticeable and important.
2) Einstein's theory of **special relativity** only works in **reference frames** that **aren't accelerating** — called **inertial frames**.
3) It's based on two assumptions:

Physical laws have the same form in all inertial frames.
The speed of light in free space is invariant.

Invariant means that it always has the same value.

4) One rather interesting consequence of special relativity is that:

No particle that has mass can move at a speed **greater than** or **equal to** the **speed of light**, c.

The **Mass** of an Object **Increases** with **Speed**

1) Special relativity is where the idea that **energy is equivalent to mass** comes from. → $E_{rest} = mc^2$
2) It means that the more you **increase** the **kinetic energy** of a mass (like a particle in an accelerator), the **more massive** it gets.
3) This happens to **any** object with kinetic energy, but it's only noticeable at **speeds approaching** c.
4) Particle accelerators have to **alter** their magnetic and electric fields to compensate for the relativistic mass of the accelerating particles.

The **Relativistic Factor** — $E_{tot} \div E_{rest}$

The **relativistic factor**, γ, depends on the speed of the particle (see page 31). It's given by:

$$\gamma = \frac{1}{\sqrt{1 - \frac{v^2}{c^2}}}$$

It's also equal to the total energy of a particle (i.e. its rest energy plus its kinetic energy) divided by its rest energy, so: → $E_{total} = \gamma E_{rest}$

For particles travelling at low speeds ($v << c$), γ will be very close to 1.

Example: An electron is accelerated to almost the speed of light. It has a rest energy of 5.1×10^5 eV. If $\gamma = 235$, find the total energy of the accelerated electron in MeV.

$E_{total} = \gamma E_{rest} = 235 \times (5.1 \times 10^5) = 1.1985 \times 10^8$ eV = **120 MeV (to 2 s.f.)**

Practice Questions

Q1 Describe the similarities and differences between a linear particle accelerator and a cyclotron.

Q2 Describe how a synchrotron works.

Q3 State what happens to the mass of a particle when it approaches the speed of light.

Q4 Write down an expression for the relativistic factor in terms of the speed of the particle and the speed of light.

Exam Questions

Q1 A proton is accelerated by a synchrotron to a total energy of 500 GeV.
Show that the relativistic factor for a proton of this energy is about 500.
($m_p = 1.673 \times 10^{-27}$ kg, 1 eV = 1.60×10^{-19} J) [3 marks]

Q2 A proton in a cyclotron is travelling at a speed of 6.82×10^5 ms^{-1}. It experiences a force of 4.91×10^{-13} N due to a perpendicular magnetic field. Calculate the strength of the magnetic flux density inside the cyclotron. [2 marks]

Smash high-energy particles together to see what they're made of...

The three types of particle accelerator all have their advantages, but the synchrotron wins hands down for generating very high energy particles. Synchrotrons get pretty big — the LHC is a whopping 27 km loop...

Electron Energy Levels

You'll probably recognise this stuff — you met it in year 1 of A level.
Electrons only exist in set energy levels and leap into higher energy levels when they get excited.

Electrons in Atoms Have Discrete Energy Levels

1) **Electrons** in an **atom** can **only exist** in certain **well-defined energy levels**. Each level is given a **number** (called the **principal quantum number** of the electron in that state), with **n = 1** representing the electron's lowest possible energy — its **ground state**.
2) The diagram on the right shows the **energy levels** for **atomic hydrogen**.
3) The **energies involved** are **so tiny** that it makes sense to use a more **appropriate unit** than the **joule**. The **electronvolt (eV)** is used instead:

> An **electronvolt** is the **kinetic energy carried** by an **electron** after it has been **accelerated** through a **potential difference** of **1 volt. 1 eV = 1.60 × 10^{-19} J.**

LEVEL | ENERGY
n = ∞ — zero energy
n = 5 — −0.54 eV
n = 4 — −0.85 eV
n = 3 — −1.5 eV
n = 2 — −3.4 eV
n = 1 — −13.6 eV

The blue arrows show 10 possible transitions between n = 5 and n = 1.

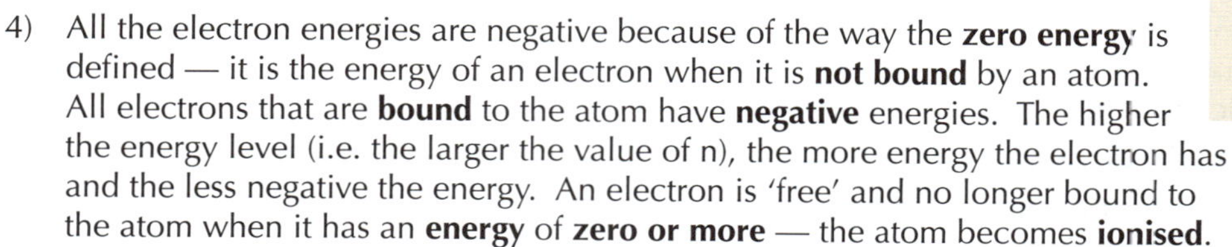

4) All the electron energies are negative because of the way the **zero energy** is defined — it is the energy of an electron when it is **not bound** by an atom. All electrons that are **bound** to the atom have **negative** energies. The higher the energy level (i.e. the larger the value of n), the more energy the electron has and the less negative the energy. An electron is 'free' and no longer bound to the atom when it has an **energy** of **zero or more** — the atom becomes **ionised**.

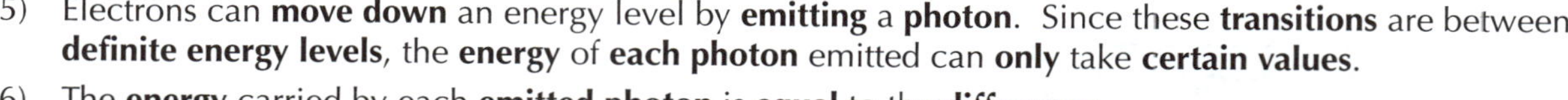

5) Electrons can **move down** an energy level by **emitting** a **photon**. Since these **transitions** are between **definite energy levels**, the **energy** of **each photon** emitted can **only** take **certain values**.
6) The **energy** carried by each **emitted photon** is **equal** to the **difference in energies** between the **two levels** that the electron has moved between. The equation on the right is for a **transition** between levels **n = 2** and **n = 1**:

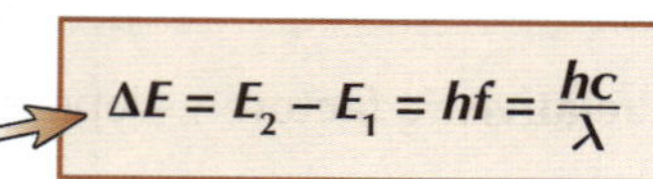

$$\Delta E = E_2 - E_1 = hf = \frac{hc}{\lambda}$$

h = Planck's constant,
f = frequency of photon,
c = speed of light in a vacuum,
λ = wavelength of photon

7) In the same way, atoms can only **absorb** allowed photon energies. When an atom **absorbs a photon**, an electron **moves up** to another energy level. This **quantisation** of electron energies in atoms produces **line emission** and **absorption spectra** (see below).

> Electrons (as well as protons and neutrons) are **fermions**. That means they obey the **Pauli exclusion principle**. This states that **no two fermions** can be in **exactly** the same **quantum state** at the same time. In the context of energy levels, that means **no more than two** electrons can be in the same **energy level** at the same time.

The Evidence for Energy Levels — Line Spectra

1) The **spectrum** of **white light** is **continuous**.
2) If you **split** the **light** up with a **prism**, the **colours** all **merge** into each other — there **aren't** any **gaps** in the spectrum.

3) You get a **line absorption spectrum** when **light** with a **continuous spectrum** passes through a cool gas.
4) At **low temperatures**, **most** of the **electrons** in the **gas atoms** will be in their **ground states**.
5) **Photons** of the **correct wavelength** are **absorbed** by the **atoms** to **excite** the **electrons** to **higher energy levels**.
6) These **wavelengths** are then **missing** from the **continuous spectrum** when the light **comes out** the other side of the gas.
7) You see a **continuous spectrum** with **dark lines** in it corresponding to the **absorbed wavelengths**.

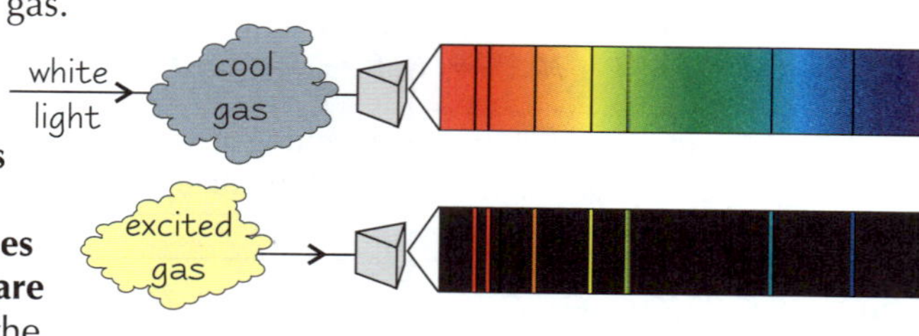

8) When an electron falls into a **lower** energy level, it **emits** a photon. **Emission spectra** show the wavelengths of photons emitted. They are made up of a series **bright lines** corresponding to the **wavelengths emitted**. If you **compare** the **absorption** and **emission spectra** of a **particular gas**,the **dark lines** in the **absorption spectrum match up** to the **bright lines** in the **emission spectrum**.

Electron Energy Levels

More Evidence — *Fluorescent Tubes*

1) Fluorescent tubes use **excited electrons** to produce light. A **high voltage** is applied across mercury vapour, which **accelerates** free electrons that **ionise** some of the mercury atoms, producing even more **free electrons**.
2) When the free electrons **collide** with electrons in other mercury atoms, the electrons in the mercury atoms are **excited** to **higher energy levels**. When they return to their ground states, they **emit UV photons**.
3) A phosphorus coating inside the tube **absorbs** these photons, in turn **exciting** its electrons to much higher orbits. These electrons then **cascade** down the energy levels, **emitting photons** in the form of **visible light**.

The *Wave Model* of the *Atom* can Help you Understand *Energy Levels*

1) Since light has both **particle** and **wave** characteristics, de Broglie suggested that **electrons** should have a **wave-like character**. Specifically, when they're in orbit **around a nucleus** they ought to behave like the **standing waves** that are formed on a guitar string when it's plucked.
2) Just as standing waves on the guitar string only exist at certain **well-defined frequencies**, only certain standing waves are possible in an atom. The **wavelength** of the electron waves should fit the **circumference** of the orbit a **whole number** of times.
3) The **principal quantum number**, **n** (corresponding to the number of the energy level), is equal to the number of **complete waves** that fit the circumference.
4) You can think of the electrons as being trapped by a **potential well** made by the nucleus. That way you can think of them as being standing waves between **two fixed walls**.
5) An electron will be able to **escape** the potential well when it has a total energy of **zero or more** (i.e. when KE + PE ≥ 0).
6) You can use the diameter of the atom, d, to find the **de Broglie wavelength**, $\lambda = \frac{h}{p} = \frac{h}{mv}$, of the electron. For n = 1, there is **half a wavelength** contained between the walls, so $\lambda = 2d$. For n = 2, there is **one complete wavelength** contained between the walls, so $\lambda = d$. For n = 3, $\lambda = \frac{2}{3}d$ and so on...

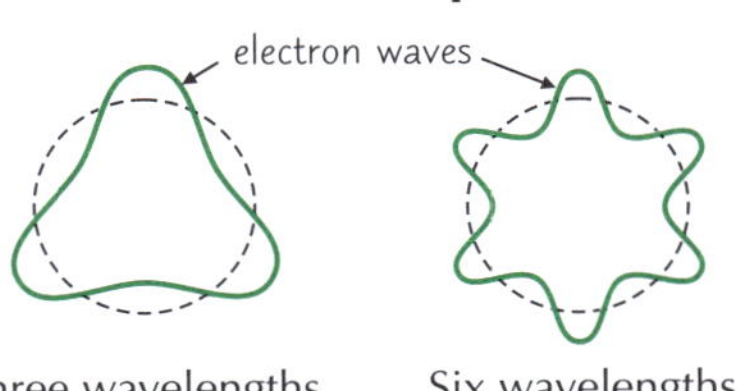

Three wavelengths n = 3 Six wavelengths n = 6

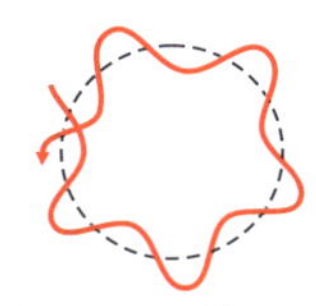
Not a standing wave. Forbidden energy.

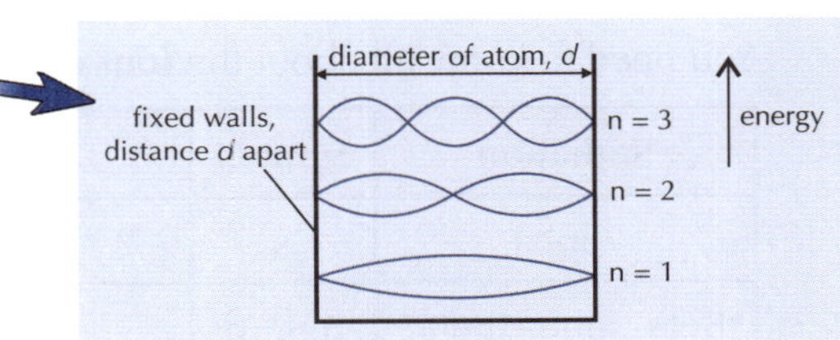

Practice Questions

Q1 Write down the equation you would use to find the difference in energy between two energy levels in an atom.

Q2 Describe how line spectra give evidence for the existence of discrete energy levels in atoms.

Q3 Describe the standing-wave model of electrons in an atom.

Exam Questions

Q1 An electron is created in beta decay and has an energy of 3.9×10^{-14} J.
State and explain whether this electron will remain bound to the nucleus. [2 marks]

Q2 The Balmer series is a series of spectral lines due to photons emitted by excited hydrogen atoms.
One Balmer line is caused by photons with a frequency of 4.57×10^{14} Hz.
a) Find the energy in joules of the photons that make up this line ($h = 6.63 \times 10^{-34}$ *Js*). [1 mark]
b) Use the diagram of the energy levels in a hydrogen atom on page 70 to identify the energy level transition that causes this spectral line. [2 marks]

Q3 The length of 1 second is defined using the radiation corresponding to a particular quantum transition in a caesium-133 atom. The difference in energy levels is 3.8×10^{-5} eV.
Calculate the frequency of this radiation. [3 marks]

My energy level's about n = 1 after that — I need chocolate...

I always find this stuff the trickiest — I mean, it's not like you can see an electron skipping about inside an atom every day, and the 'electrons are waves as well' plot twist keeps you on your toes. Once you've been through it all a few times though it starts to click, so stick with it and soon it'll be as easy as an electron transition from n = 2 to n = 1.

Radioactive Emissions

Make sure you're happy with the previous section before you start this, or it might not make much sense...

Atomic Structure can be Represented Using Standard Notation

Standard notation summarises all the information you need about an element's **atomic structure**:

$^{12}_{6}C$

The **proton number** or **atomic number** (*Z*) — there are six protons in a carbon atom.

The **nucleon number** or **mass number** (*A*) — there are a total of 12 protons and neutrons in a carbon-12 atom.

The symbol for the element. For carbon, it's C.

Atoms of the **same element** (i.e. that have the same number of protons) that have different numbers of **neutrons** (same *Z*, different *A*) are called **isotopes**.

Unstable Isotopes are Radioactive

1) The nucleus is under the **influence** of the **strong nuclear force holding** it **together** and the **electrostatic force pushing** the **protons apart**. It's a very **delicate balance**, and it's easy for a nucleus to become **unstable**.
2) If a nucleus is **unstable**, it **transforms** into a more stable isotope by emitting **radiation**. Its **instability** could be caused by:
   - **too many neutrons**
   - **too few neutrons**
   - **too many nucleons** in total (it's **too massive**)
   - **too much energy** in the nucleus
3) The nuclei of some isotopes are unstable and decay — we say these isotopes are **radioactive**.

There are Four Types of Nuclear Radiation

You need to know all about the **four** different types of **nuclear radiation** — here's a handy **table** to get you started.

Radiation	Symbol	Constituent	Relative Charge	Mass (u)
Alpha	α	2 protons and 2 neutrons (a helium nucleus)	+2	4
Beta-minus (Beta)	β or β^-	Electron	−1	Negligible
Beta-plus	β^+	Positron (see p.63)	+1	Negligible
Gamma	γ	Short-wavelength, high-frequency EM wave.	0	0

The masses in this table are given in **atomic mass units** (u).
An atomic mass unit is roughly equal to the mass of a proton or neutron.

The Different Types of Radiation have Different Penetrations

When radiation **hits** an **atom** it can **knock off electrons**, creating an **ion** — so, **radioactive emissions** are also known as **ionising radiation**. The **different types** of radiation have **different ionising powers** as well as different **speeds** and **penetrating powers**.

Radiation	Symbol	Ionising Power	Speed	Penetrating power
Alpha	α	Strong	Slow	Absorbed by paper or a few cm of air
Beta-minus (Beta)	β or β^-	Weak	Fast	Absorbed by ~3 mm of aluminium
Beta-plus	β^+	Annihilated by electron — so virtually zero range		
Gamma	γ	Very weak	Speed of light	Absorbed by many cm of lead, or several m of concrete

α and β particles are also affected by magnetic fields, as they carry a charge. γ radiation isn't affected by a magnetic field. There's more on charged particles in magnetic fields on p.58.

The penetrating power of radiation **decreases** with increasing ionising power. This is because radiation loses energy as it ionises atoms. This means that the **higher** the **ionising power** of a type of radiation, the **more energy** it **loses** in a given distance, so the **shorter** its **range**.

Radioactive Emissions

You can *Investigate* the *Penetration* of Different Kinds of Radiation in the Lab

You can investigate the penetration of different kinds of radiation by using different radioactive sources. These can be **dangerous** if you don't use them properly:

- Radioactive sources should be kept in a **lead-lined box** when they're not being used.
- They should only be picked up using **long-handled tongs** or **forceps**.
- Take care not to **point** them at anyone.

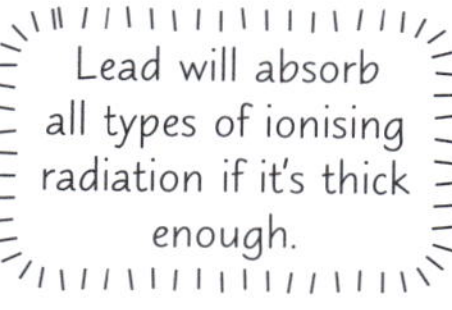

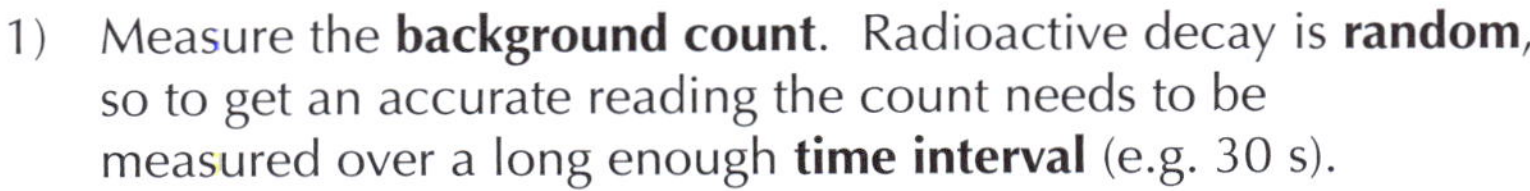

1) Measure the **background count**. Radioactive decay is **random**, so to get an accurate reading the count needs to be measured over a long enough **time interval** (e.g. 30 s).

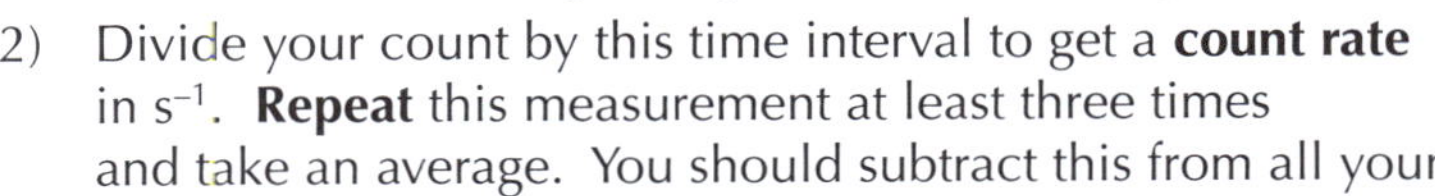

2) Divide your count by this time interval to get a **count rate** in s^{-1}. **Repeat** this measurement at least three times and take an average. You should subtract this from all your results.

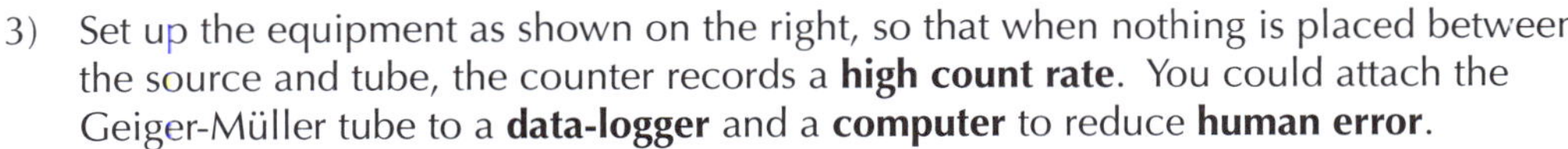

3) Set up the equipment as shown on the right, so that when nothing is placed between the source and tube, the counter records a **high count rate**. You could attach the Geiger-Müller tube to a **data-logger** and a **computer** to reduce **human error**.

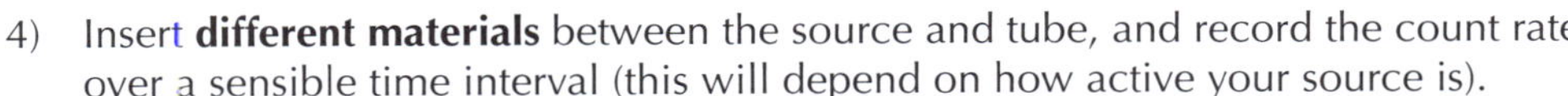

4) Insert **different materials** between the source and tube, and record the count rate over a sensible time interval (this will depend on how active your source is).

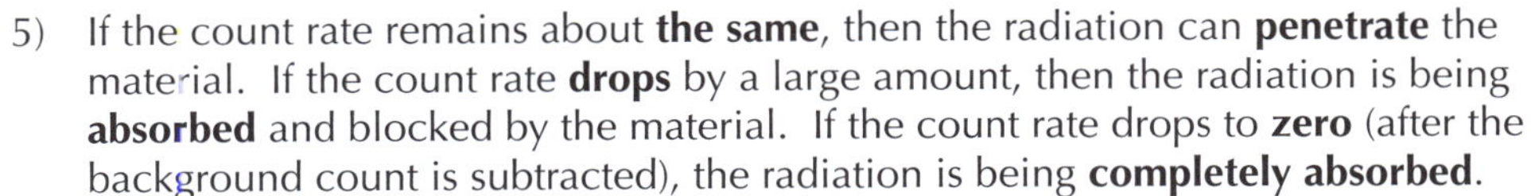

5) If the count rate remains about **the same**, then the radiation can **penetrate** the material. If the count rate **drops** by a large amount, then the radiation is being **absorbed** and blocked by the material. If the count rate drops to **zero** (after the background count is subtracted), the radiation is being **completely absorbed**.

If you're comparing penetration across different materials, they're unlikely to all have the same thickness. Bear this in mind when you draw your conclusions.

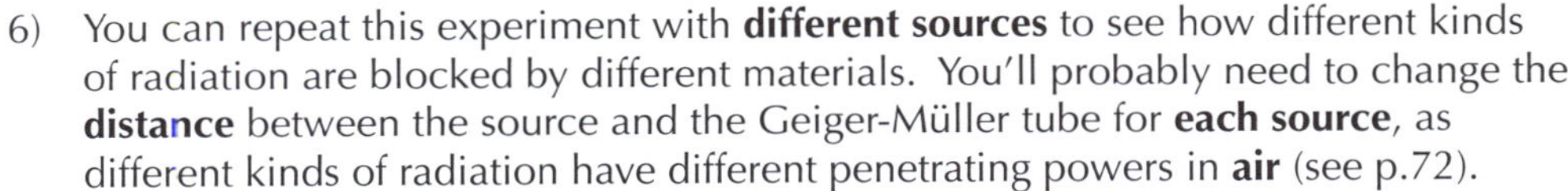

6) You can repeat this experiment with **different sources** to see how different kinds of radiation are blocked by different materials. You'll probably need to change the **distance** between the source and the Geiger-Müller tube for **each source**, as different kinds of radiation have different penetrating powers in **air** (see p.72).

7) You could also investigate how the count rate from a particular source is affected by the **thickness** of a given blocking material, e.g. by using **multiple sheets** of **aluminium** for **beta** radiation, or different thicknesses of **lead** for **gamma** radiation.

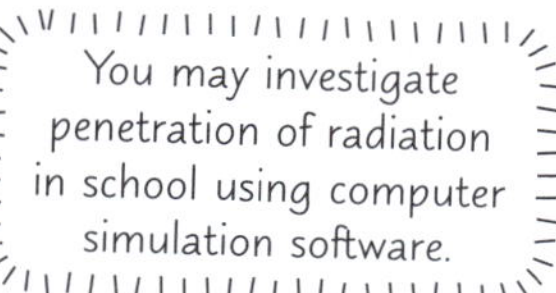

You Can *Represent Nuclear Decay* Using *Equations*

1) You usually write the particles involved in nuclear decay in **standard notation** (p.72) so you can see what happens to the **protons** and **neutrons**. E.g., the decay of americium-241 to neptunium-237 looks like this:

$${}^{241}_{95}\text{Am} \longrightarrow {}^{237}_{93}\text{Np} + {}^{4}_{2}\alpha$$

2) The nucleus you start with is called the **parent nucleus** (here it's americium-241), and the nucleus it decays to is called the **daughter nucleus** (here it's neptunium-237).

3) Decay equations need to be **balanced** — in every nuclear reaction, including fission and fusion (p.80–81), **charge**, **nucleon number** and **lepton number** (p.62) must be **conserved**.

4) **Beta-minus** particles have a **negative** charge. They're written with a **negative proton number** (${}^{0}_{-1}\beta$).

5) **Energy** and **momentum** are also **conserved** in all nuclear reactions. Mass, however, **doesn't** have to be **conserved** — the **mass** of an alpha particle is **less** than the **individual masses** of **two protons** and **two neutrons**. The difference in mass is called the **mass defect**, and the **energy released** when the nucleons **bond together** to form the alpha particle accounts for the missing mass. More on that later (p.78).

α *Emission* Happens in *Heavy Nuclei*

Alpha emission only happens in **very heavy** atoms, like **uranium** and **radium**. The **nuclei** of these atoms are **too massive** to be stable.

When an alpha particle is emitted, the **proton number decreases** by **two**, and the **nucleon number decreases** by **four**.

Example:

$238 = 234 + 4$ — nucleon numbers balance

$${}^{238}_{92}\text{U} \longrightarrow {}^{234}_{90}\text{Th} + {}^{4}_{2}\alpha$$

$92 = 90 + 2$ — proton numbers balance, so charge is conserved

Radioactive Emissions

β^- Emission Happens in Neutron-Rich Nuclei

1) **Beta-minus** decay is the emission of an **electron** from the **nucleus** along with an **antineutrino** (p.63). It happens in isotopes that are **'neutron rich'** (have many more **neutrons** than **protons** in their nuclei).
2) One of the **neutrons** in the nucleus **decays** into a **proton** and ejects a beta particle and an antineutrino.

When a beta-minus particle is emitted, the **proton number increases** by **one**, and the **nucleon number stays the same**.

Example: nucleon numbers balance

$$188 = 188 + 0 + 0$$

$${}^{188}_{75}Re \longrightarrow {}^{188}_{76}Os + {}^{0}_{-1}\beta + {}^{0}_{0}\bar{\nu}$$

$$75 = 76 - 1 + 0$$

proton numbers balance, charge is conserved

β^+ Emission Happens in Proton-Rich Nuclei

In **beta-plus emission**, a **proton** gets **changed** into a **neutron**, releasing a **positron** and a **neutrino**.

When a beta-plus particle is emitted, the **proton number decreases** by **one**, and the **nucleon number stays the same**.

Example:

$$18 = 18 + 0 + 0$$ — nucleon numbers balance

$${}^{18}_{9}F \longrightarrow {}^{18}_{8}O + {}^{0}_{+1}\bar{\beta} + {}^{0}_{0}\nu$$

$$9 = 8 + 1 + 0$$ — proton numbers balance

In both types of beta decay, a lepton and an anti-lepton are produced, so lepton number is conserved (see page 62). In β^- decay it's an electron and an antineutrino, and in β^+ decay, it's a positron and a neutrino.

γ Radiation is Emitted from Nuclei with Too Much Energy

Gamma rays can be emitted from a nucleus with **excess energy** — we say the nucleus is **excited**. This energy is **lost** by emitting a **gamma ray**. This often happens after an **alpha** or **beta** decay has occurred.

During **gamma emission**, there is **no change** to the nuclear **constituents** — the nucleus just **loses excess energy**.

Practice Questions

Q1 What is an isotope?

Q2 List four factors that could make a nucleus unstable.

Q3 What are the four types of nuclear radiation? What does each one consist of?

Q4 What is one atomic mass unit approximately equal to?

Q5 Which type of radiation has the greatest penetrating power? Which is the most strongly ionising?

Q6 Describe the changes that happen in a nucleus during alpha, beta-minus, beta-plus and gamma decay.

Exam Questions

Q1* A source is known to emit a single type of radiation (alpha, beta-minus or gamma). Describe an experiment to identify the type of radiation that the source emits. [6 marks]

Q2 a) Radium-226 (Ra, proton number 88) decays to radon (Rn) by emitting an alpha particle. Write a balanced nuclear equation for this reaction. [3 marks]

b) Potassium-40 (K, proton number 19) decays to calcium (Ca) by emitting an electron. Write a balanced nuclear equation for this reaction. [3 marks]

* The quality of your extended response will be assessed in this question.

Radioactive emissions — as easy as α, β, γ...

You need to learn the different types of nuclear radiation and their properties, and how to write equations for all the different types of decay. Just remember that charge and nucleon number are conserved, and you won't go far wrong...

Dangers of Radiation

When you work with nuclear radiation, you need to carefully weigh up the risks and the benefits.

Using Radioactive Materials has Benefits and Risks

1) **Radioactive materials** can be useful — for example, they're used to **generate power** (p.80-81), in **medicine** for **diagnosis** and **treatment,** and to **kill harmful microorganisms** that might contaminate our **food**.
2) However, they're also dangerous. They can cause **cancerous tumours**, **skin burns**, **sterility**, **radiation sickness**, **hair loss** and even **death**.
3) The result is that radiation is only used when the benefits outweigh the **risks**. There are two parts to the risk: **how likely** it is that the radiation will cause a problem, and **how bad** the problem would be if it did happen.

For example:

1) A **nuclear reactor** melting down would be **catastrophic**, but it's also **very unlikely**, so the risk might be acceptable.
2) Ionising radiation can **cause** cancer, but it can also be used in cancer **treatments** to destroy tumours — see below. The risk of serious damage caused by the treatment is considered **acceptable** if the treatment is likely to **prolong** or **improve** a patient's life.

You may see this idea expressed as an equation, in the form: risk = probability × consequences.

4) We need to be able to **measure** the dangers radiation poses to humans to make sensible decisions about safety.

Radiation can Cause a lot of Harm to Body Tissues

1) Ionising radiation gets its name because it can knock electrons off atoms, creating ions. When this happens to cells in your body, it can **damage** cells, cause them to **mutate** or even **kill them**.
2) The amount of **energy deposited per kilogram** of tissue is called the **absorbed dose**, and is measured in **grays** (**Gy**).
3) To calculate an absorbed dose, you first need the energy of the radiation **absorbed** by a mass of tissue.

$$\textbf{absorbed dose (Gy)} = \frac{\textbf{energy deposited (J)}}{\textbf{mass (kg)}}$$

4) You can find the energy **released** by a source from its activity, the energy of the radiation, and the time interval. If all of the released energy is absorbed by tissue, you can calculate the absorbed dose using the tissue's mass.

Example: A Pu-238 source (half life 88 years) releases alpha particles with an average energy of 8.78×10^{-13} J. Its activity is 1.55×10^4 Bq. Proper safety procedures are not followed in a lab, and a scientist holds the source with his bare hand. Assuming all of the radiation emitted is absorbed by 0.05 g of tissue, calculate the absorbed dose the scientist will receive if he holds the source for ten seconds.

As the half-life of the source is so long, you can assume the activity remains constant during the 10 second period.

1 Bq = 1 decay per second, so multiply the activity by ten to get the number of decays in ten seconds: $(1.55 \times 10^4) \times 10 = 1.55 \times 10^5$

Each decay releases on average 8.78×10^{-13} J. So the energy released in ten seconds is: $(8.78 \times 10^{-13}) \times (1.55 \times 10^5) = 1.3609 \times 10^{-7}$ J

You're told the energy deposited is equal to the energy emitted by the source, so Absorbed dose = energy deposited ÷ mass: $(1.3609 \times 10^{-7}) \div (0.05 \times 10^{-3}) = 2.7218 \times 10^{-3}$ $= \mathbf{2.7 \times 10^{-3}}$ **Gy (to 2 s.f.)**

Radiation can be Used to Treat Cancer

1) The ability of radiation to damage or kill cells makes it a useful tool to treat **cancer**. However, it's important when treating cancer with radiation to **minimise** the dose absorbed by **healthy tissue** at the same time as **maximising** the dose absorbed by the cancer cells.
2) One way of doing this is to place radioactive sources **inside** the patient, in or next to the tumour.
3) Sources can either be inserted next to the tumour for a **short period** of time, or left within the patient **permanently**.
4) Sources left within a patient permanently are chosen to deliver a **lower dose** per hour than sources inserted for a shorter duration, to **minimise damage** to healthy tissue. They will also have a fairly **short half-life** (p.6) — around 20 or so days. This means they remain active for **long enough** for the cancer cells to be killed (typically a few months) without causing unnecessary damage to surrounding tissue.

Dangers of Radiation

The Quality Factor Affects How Much Damage is Done

1) The amount of **tissue damage** caused by exposure to radiation isn't just due to the amount of energy absorbed — it also depends on the **type of ionising radiation** and the **type of body tissue**.
2) The **effective dose** is a measure that lets you **compare the amount of damage** to body tissues that have been **exposed** to different types of radiation:

Effective dose = absorbed dose × quality factor

The quality factor is sometimes called the radiation quality factor.

3) The **unit** of effective dose is the **sievert** (**Sv**).
4) The table on the right shows typical values for the **quality factor**.
For example, if you exposed a sample of **body tissue** to **1 Gy** of **alpha** radiation, it could do the **same damage** as an exposure of 20 Gy of **gamma** radiation on the same type of body tissue.

Radiation	Typical quality factor	Effective dose of 1 Gy
alpha	20	20 Sv
beta	1	1 Sv
gamma	1	1 Sv

Alpha and Beta Particles have Different Ionising Properties

The higher a radiation's quality factor, the more **ionising** it is. Alpha particles are the most strongly ionising form of radiation, followed by beta particles, then gamma radiation (p.72).

Remember, ionisation happens when electrons are pulled from, or knocked off, an atom.

1) When **beta** or **alpha** particles ionise an atom, they **transfer** some of their **energy** to the atom being ionised.
2) **Alpha** particles are **strongly positive** — so they can **easily pull electrons** off atoms.
3) This means an alpha particle **quickly ionises** many atoms (about 10 000 ionisations per alpha particle) and **loses** all its **energy** — that's why it causes so much **damage** to body tissue.
4) The **beta**-minus particle has **lower mass** and **charge** than the alpha particle, but a **higher speed**. This means it can still **knock electrons** off atoms. Each **beta** particle will ionise about 100 atoms, **losing energy** at each interaction.
5) This **lower** number of **interactions** means that beta radiation causes much **less damage** to body tissue than alpha radiation — explaining the **lower radiation quality factor**.

The Intensity of Gamma Radiation Decreases with Distance

1) Gamma radiation is the most **weakly ionising** form of nuclear radiation, but also the most **penetrating** (see page 72).
2) This means that, although its quality factor is generally **low**, it is more difficult to **shield** yourself from — you'd need many centimetres of lead to block all the radiation coming from a gamma source. This needs to be taken into account when thinking about **lab safety**.
3) When gamma radiation travels through an **absorbing material** (e.g. concrete), its **intensity** (the amount of **radiation per unit area**) **decreases exponentially**. So people working near gamma sources with high activity can be protected by the use of thick lead or concrete shielding.
4) A **gamma source** will also **emit** gamma **radiation** in **all directions**. However, this radiation **spreads out** as it gets **further away** from the source, so the **intensity** will **decrease** the further you get from the source. This means you can also protect yourself from gamma sources by simply keeping your **distance**.

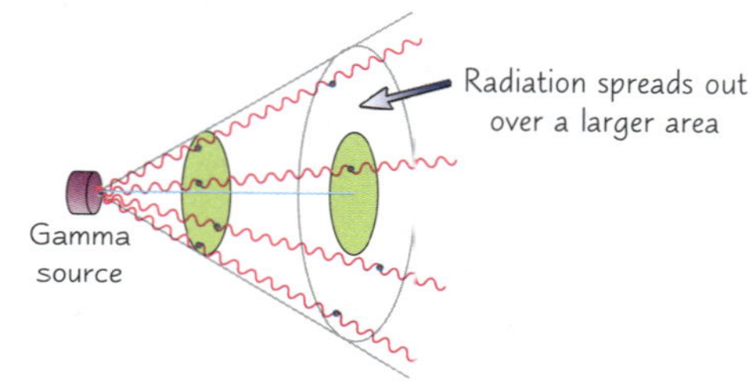

It's important to keep your distance from all radioactive sources that you work with, and you should always follow the safety measures given on page 73.

Dangers of Radiation

You Need to Consider the Half-Life when Thinking About Risk

1) The activity of a radioactive source **decreases** over time.
2) As you know from page 5, the **half-life**, $T_{1/2}$, of a radioactive source is the time it takes for the number of **radioactive** nuclei in the source to **halve**. This means it's also the amount of time it takes for the **activity** of the source to halve. You can find the half-life of a radioactive isotope like protactinium-234 experimentally using the method on page 5.
3) The half-life of an isotope is related to its **decay constant**, λ, by the formula:

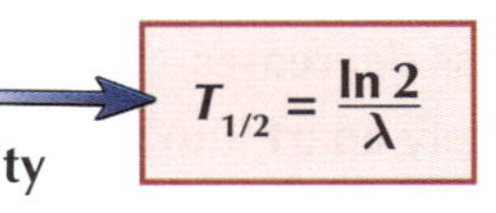

$$T_{1/2} = \frac{\ln 2}{\lambda}$$

Where ln is the natural log (p.84), λ is the decay constant in s^{-1}, and $T_{1/2}$ is the half-life in seconds.

4) You can calculate a radioactive source's **activity** (the number of decays per second) from its decay constant using the formula:

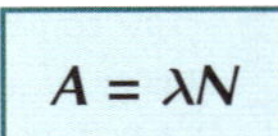

$$A = \lambda N$$

A is the activity in Bq, and *N* is the number of undecayed nuclei in the source.

If this stuff seems unfamiliar, flip back to pages 4-6 for a recap.

5) We need to consider $T_{1/2}$ when assessing the **dangers** of a radioactive source and planning how to manage them:

For example:
1) **Radioactive waste** often has a very **long half-life**, which needs to be considered when deciding whether or not to use nuclear power, and deciding where nuclear waste should be **stored**.
2) Radioactive isotopes used as **medical tracers** need to have a **short half-life** to minimise the exposure of patients to radiation, and the chance of environmental contamination when the body excretes them.

Tracers are radioactive sources that are ingested or injected into patients. The radiation they release as they move through the body can then be used to generate images that are useful for medical diagnostics.

Practice Questions

Q1 What factors do we need to take into account when trying to measure the risk posed by radioactive materials?

Q2 What is the difference between 'absorbed dose' and 'effective dose'?

Q3 Describe how the intensity of gamma radiation changes with distance from the source.

Q4 State the equation for calculating the activity of a radioactive source from its decay constant.

Exam Questions

Q1 Radioactive sources need to be handled carefully to prevent harm to users.

a) State two health risks radioactive sources pose to humans. [2 marks]

b) Calculate which of the following absorbed doses would cause the greatest damage to a sample of body tissue: 0.6 Gy of alpha radiation with a quality factor of 20, or 9 Gy of beta radiation with a quality factor of 1. [2 marks]

Q2 A patient is being treated for cancer. The treatment involves placing a radioactive source within the tumour, consisting of an isotope with a decay constant of 3.21×10^{-7} s^{-1}. Each decay produces gamma radiation with an average energy of 2.9×10^{-15} J per decay. The source's activity is 5.03×10^{8} Bq. The tumour has a mass of 4.2 g.

a) Calculate the half-life of the isotope. [1 mark]

b) Assuming that 40% of the radiation emitted by the source is deposited in the tumour, calculate the absorbed dose delivered to the tumour per second, in grays. [2 marks]

c) Calculate how long it would take for the source to deliver a dose of 0.50 Gy to the tumour. [1 mark]

It's all about risks, benefits, probabilities and consequences...

Radioactive sources can be dangerous, but they can do a lot of good too. We need to weigh up the risks and the benefits carefully when we're using radiation, and (as always) it's important to take safety procedures seriously.

Binding Energy

Turn off the radio and close the door, 'cos you're going to need to concentrate hard on this stuff about binding energy...

The Mass Defect is Equivalent to the Binding Energy

1) The **mass** of a **nucleus** is **less than** the mass of its **constituent parts** — the difference is called the **mass defect**.
2) Mass and energy are **equivalent**, according to Einstein's equation: (there's more on this equation on pages 64-65).

$$E_{rest} = mc^2$$

E_{rest} is the rest energy in J, m is the mass in kg and c is the speed of light in a vacuum in ms^{-1}.

3) As nucleons join together, the total mass **decreases** — this '**lost**' mass is **converted** into energy and **released**.
4) If you then **pulled** the nucleus completely **apart** into its separate nucleons, the **energy** you'd have to use to do it would be the **same** as the energy **released** when the nucleus formed. This is called the **binding energy** and is usually measured in MeV. The binding energy is **equivalent** to the **mass defect**.
5) The binding energy and the mass defect are both **negative** quantities. That's because they correspond to mass (and thus energy) **lost** from the nucleus.
6) **Calculating** the binding energy from the mass defect is **pretty easy** — it's just $E = mc^2$, where m is the mass defect:

You could also find the binding energy of a nucleus by calculating the rest energy of the nucleus and the rest energy of its constituent parts using the equation above and finding the difference.

Example: Calculate the binding energy (in MeV) of the nucleus of a lithium atom, $^{6}_{3}Li$, given that its mass defect is –0.0343 u.

1) Convert the mass defect into kg:

$$\text{Mass defect} = -0.0343 \times (1.661 \times 10^{-27}) = -5.69723 \times 10^{-29}\ \text{kg}$$

2) Use $E = mc^2$ to calculate the binding energy:

$$E = (-5.69723 \times 10^{-29}) \times (3.00 \times 10^{8})^2 = -5.127507 \times 10^{-12}\ \text{J}$$

3) Convert your answer into electron volts:

$$(-5.127507 \times 10^{-12}) \div (1.60 \times 10^{-19}) = -3.204... \times 10^{7}\ \text{eV}$$

$1\ \text{eV} = 1.60 \times 10^{-19}\ \text{J}$ $\quad$ **= –32.0 MeV (to 3 s.f.)**

Atomic mass is usually given in atomic mass units (u), where $1\ \text{u} = 1.661 \times 10^{-27}$ kg (see page 72).

The Binding Energy Per Nucleon is at a Maximum around N = 50

A useful way of **comparing** the binding energies of different nuclei is to look at the **binding energy per nucleon**.

$$\textbf{Binding energy per nucleon (in MeV)} = \frac{\textbf{Binding energy } (B)}{\textbf{Nucleon number } (A)}$$

So, the binding energy per nucleon for $^{6}_{3}Li$ (in the example above) is –32 ÷ 6 = –5.3 MeV (to 2 s.f.).

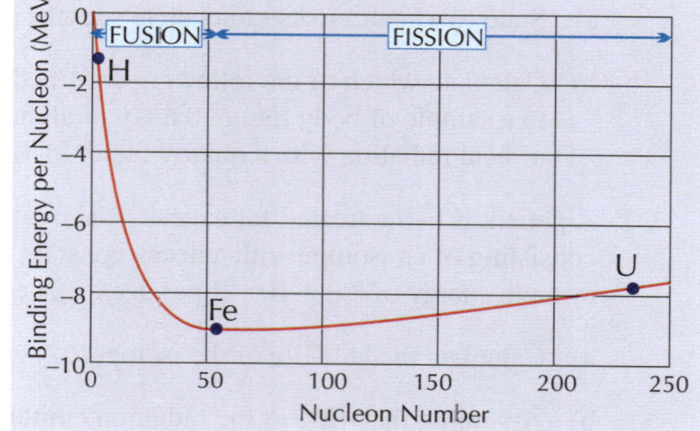

1) If you plot a **graph** of **binding energy per nucleon** against **nucleon number**, for all elements, the line of best fit shows a **curve** — sometimes called a **nuclear valley**.
2) The more negative the binding energy per nucleon, the **more energy** is needed to **remove** nucleons from the nucleus — so the more stable the nucleus.
3) The **most stable** nuclei occur around the **minimum point** on the graph — which is at **nucleon number 56** (i.e. **iron**, Fe). Nuclei with a nucleon number close to 56 are bound **most strongly**.
4) Generally, the nuclear reactions that tend to happen are those that make nuclei **more stable** (apart from in extreme conditions such as during a supernova), so nuclei undergoing nuclear reactions will tend to move towards this valley of stability.
5) **Combining small nuclei** is called nuclear **fusion** (p.81) — this makes the **binding energy per nucleon** much more negative, which means a lot of **energy is released** during nuclear fusion.
6) **Fission** is when **large nuclei** are **split in two** (see p.80) — the **nucleon numbers** of the two **new nuclei** are **smaller** than the original nucleus. This makes the binding energy per nucleon **more negative**. So, energy is also **released** during nuclear fission (but not as much energy per nucleon as in nuclear fusion).

Binding Energy

Use Binding Energies to Calculate the Energy Released in Nuclear Reactions

1) You can **calculate** the energy released in a **nuclear fission** or **nuclear fusion** reaction (p.80–81) from the **binding energies** of the parent and daughter nuclei — the energy released is equal to the difference between the binding energy of the parent nucleus (or nuclei) and the daughter nucleus (or nuclei).

Example: Uranium-235 (binding energy –1800 MeV) decays to rubidium-92 (binding energy –790 MeV) and caesium-140 (binding energy –1200 MeV) according to the equation:
$^{235}_{92}U \rightarrow ^{92}_{37}Rb + ^{140}_{55}Cs + 3\,^{1}_{0}n$ + energy. Calculate the energy released in this reaction.

Just find the change in binding energy: –1800 – (–790) – (–1200) = **190 MeV**

The neutrons aren't bound to anything, so their binding energies are zero.

2) You may have to calculate the binding energies of the nuclei involved in a reaction from average **binding energies per nucleon** (as on the graph on page 78).

Example: Deuterium and tritium can fuse to form helium, according to the equation:
$^{2}_{1}H + ^{3}_{1}H \rightarrow ^{4}_{2}He + ^{1}_{0}n$ + energy. Calculate how much energy is released in this reaction.

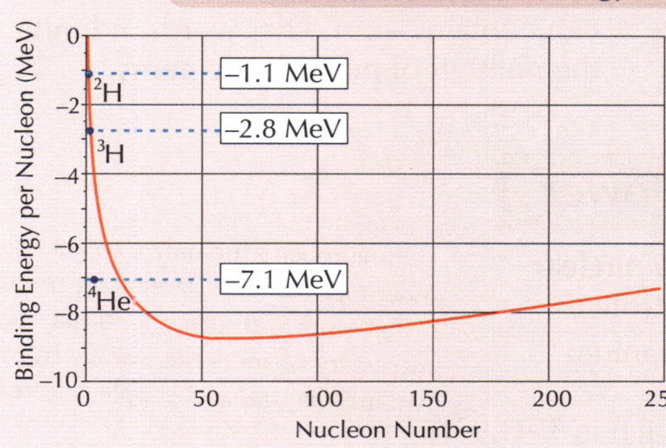

The binding energy per nucleon for deuterium is –1.1 MeV.
The binding energy per nucleon for tritium is –2.8 MeV.
The binding energy per nucleon for helium is –7.1 MeV.

Multiply each of these values by the number of nucleons in each nucleus to find the binding energy of each nucleus:

$(2 \times -1.1) + (3 \times -2.8) \rightarrow (4 \times -7.1) + 0 + \text{energy}$

$(-2.2) + (-8.4) \rightarrow (-28.4) + 0 + \text{energy}$

$(-10.6) \rightarrow (-28.4) + 0 + \text{energy}$

So, energy released = –10.6 – (–28.4) = 17.8 = **18 MeV (to 2 s.f.)**

3) Alternatively, you can calculate the energy released in a nuclear reaction by calculating the **difference in mass** between the parent nucleus (or nuclei) and the daughter nucleus (or nuclei), then using $E = mc^2$ to find the energy that this change in mass is equivalent to.

Practice Questions

Q1 What is the binding energy of a nucleus? How does it relate to the mass defect?

Q2 Sketch a graph of binding energy per nucleon against nucleon number.
Label the regions where fusion and fission occur.

Q3 Which element is the most stable? Explain your answer in terms of the binding energy per nucleon.

Exam Questions

Q1 The mass of a $^{14}_{6}C$ nucleus is 13.99995 u. The mass of a proton is 1.00728 u, and a neutron is 1.00867 u.

a) Calculate the mass defect of a $^{14}_{6}C$ nucleus. Give your answer in atomic mass units. [2 marks]

b) Calculate the binding energy of the nucleus in MeV. [3 marks]

Q2 The following equation shows a nuclear reaction between two deuterium ($^{2}_{1}H$) nuclei, to form helium-3 ($^{3}_{2}He$):

$$^{2}_{1}H + ^{2}_{1}H \rightarrow ^{3}_{2}He + ^{1}_{0}n + \text{energy}$$

a) What type of nuclear reaction is this? [1 mark]

b) The binding energy per nucleon is 0 MeV for a neutron, approximately –1.12 MeV for a $^{2}_{1}H$ nucleus, and approximately –2.57 MeV for a $^{3}_{2}He$ nucleus.
Use these values to calculate the energy released by this reaction. [3 marks]

Don't tie yourself up in knots...

This stuff is a bit of a headache — the idea of a particle having a smaller mass than the particles inside it confuses me no end — but you need to know it. You know the drill by now, back to the top of page 78, and read it all again...

Nuclear Fission and Fusion

What did the nuclear scientist have for her tea? Fission chips... hohoho.

Fission** Means **Splitting Up** into **Smaller Parts

1) **Heavy nuclei** (e.g. uranium) are **unstable** and some can randomly **split** into two **smaller** nuclei (and sometimes several neutrons) — this is called **nuclear fission**.
2) This process is called **spontaneous** if it just happens **by itself**, or **induced** if we **encourage** it to happen.

Example:

Fission can be induced by making a neutron enter a ^{235}U nucleus, causing it to become very unstable.

$^{1}_{0}n$ $^{235}_{92}U$ fission $^{92}_{36}Kr$ $^{1}_{0}n$ $^{1}_{0}n$ $^{1}_{0}n$ Energy $^{141}_{56}Ba$

Only low energy neutrons can be captured in this way. A low energy neutron is called a **thermal neutron**.

3) **Energy is released** during nuclear fission because the new, smaller nuclei have a **higher binding energy per nucleon** (see p.78) and a lower total mass.
4) The **larger** the nucleus, the more **unstable** it will be — so large nuclei are **more likely** to **spontaneously fission**.
5) This means that spontaneous fission **limits** the **number of nucleons** that a nucleus can contain — in other words, it **limits** the number of **possible elements**.

*Controlled **Nuclear Reactors** Produce Useful **Power***

We can **harness** the **energy** released by nuclear **fission reactions** in a **nuclear reactor**, but it's important that these reactions are very **carefully controlled**.

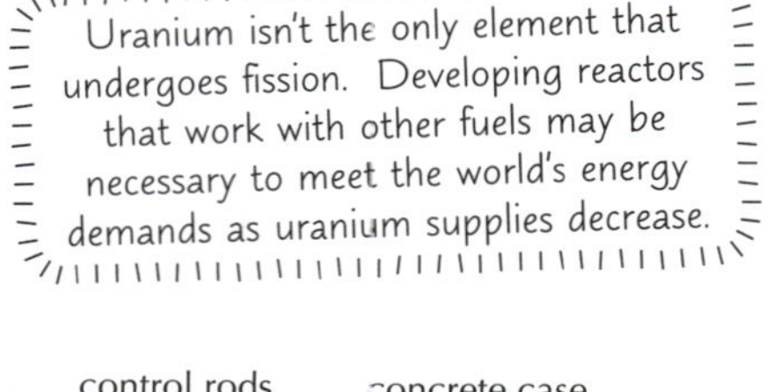
Uranium isn't the only element that undergoes fission. Developing reactors that work with other fuels may be necessary to meet the world's energy demands as uranium supplies decrease.

1) **Uranium** is the main element used as '**fuel**' in nuclear fission (uranium nuclear reactors also release some energy through the fission of plutonium). Nuclear reactors use **fuel rods of uranium** that are rich in ^{235}**U**. (The rods also contain a lot of ^{238}**U**, but that doesn't undergo fission.)
2) The **fission** of uranium produces more **neutrons** which then **induce** other nuclei to fission — this is called a **chain reaction**.
3) The **neutrons** will only cause a chain reaction if they are **slowed down** to **thermal neutron** energy levels (see above), so they can be **captured** by the uranium nuclei.
4) To do this, ^{235}U **fuel rods** need to be placed in a **moderator** (for example, **water**). You need to choose a moderator that will slow down some neutrons enough so they can cause **further fission**, keeping the reaction going at a steady rate.

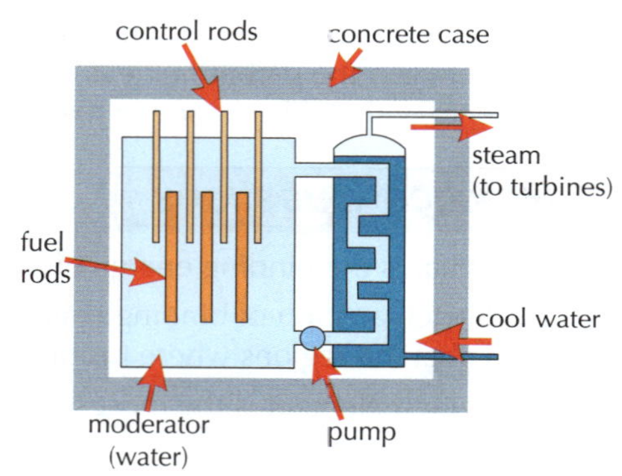

5) You want the chain reaction to continue on its own at a **steady rate**, where **one** fission follows another. This means that for each nuclei that fissions, exactly one of the neutrons released triggers another fission. The amount of 'fuel' you need to do this is called the **critical mass** — any less than the critical mass (**sub-critical mass**) and too few neutrons will be captured by other nuclei and the reaction will just peter out. Nuclear reactors use a **supercritical** mass of fuel (where several new fissions normally follow each fission) and **control the rate of fission** using **control rods**.
6) Control rods control the **chain reaction** by **limiting** the number of **neutrons** in the reactor. They **absorb neutrons** so that the **rate of fission** is controlled. **Control rods** are made up of a material that **absorbs neutrons** (e.g. boron), and they can be inserted by varying amounts to control the reaction rate.
In an **emergency**, the reactor will be **shut down** automatically by the **release of the control rods** into the reactor, which will stop the reaction as quickly as possible.
7) **Coolant** is sent around the reactor to **remove heat** produced in the fission — often the coolant is the **same water** that is being used in the reactor as a **moderator**. The **heat** from the reactor can then be used to make **steam** for powering **electricity-generating turbines**.

If the chain reaction in a nuclear reactor isn't **effectively managed** (such that too few neutrons are absorbed by the control rods), large amounts of **energy** are **released** in a very **short time**. **Many new fissions** will follow each fission, causing a **runaway reaction** which could lead to an **explosion**. This is what happens in a **fission** (**atomic**) **bomb**.

Nuclear Fission and Fusion

There are Costs and Benefits to Nuclear Fission Power Plants

1) Deciding whether or not to build a **nuclear power station** (and if so **where** to build it) is a tricky business.
2) Nuclear fission doesn't produce carbon dioxide, unlike burning **fossil fuels**, so it doesn't contribute to **global warming**. It also provides a **continuous energy supply**, unlike many renewable sources (e.g. wind/solar).
3) However, some of the **waste products** of **nuclear fission** are **highly radioactive** and difficult to handle and store.
4) When waste material is removed from the reactor, it is initially **very hot**, so it is placed in **cooling ponds** until the **temperature falls** to a safe level. The radioactive waste is then **stored** in **sealed containers** in specialist facilities until its **activity has fallen** sufficiently. This can take **many years**, and there's a risk that material could escape from these containers. A leak of radioactive material could be **harmful** to the environment and local human populations both now and in the future, particularly if the material were to contaminate **water supplies**.
5) **Accidents** or **natural disasters** pose a risk to nuclear reactors. In 2011 an earthquake and subsequent tsunami in Japan caused a meltdown at the Fukushima nuclear power plant. Over **100 000 people** were evacuated from the area, and many tonnes of contaminated water leaked into the sea. The **perceived risk** of this kind of disaster leads many people to oppose the construction of nuclear power plants near their homes.
6) Because of all of the safety precautions necessary, **building** and **decommissioning** nuclear power plants is very **time-consuming** and **expensive**.

Fusion Means Joining Nuclei Together

1) **Two light nuclei** can **combine** to create a larger nucleus. This is called **nuclear fusion.**
2) Nuclei can **only fuse** if they have enough energy to overcome the **electrostatic** (Coulomb) **repulsion** between them (p.72), and get close enough for the **strong interaction** to bind them.
3) This means fusion reactions require **much higher temperatures** than fission, as well as high pressures (or high densities). Under such conditions, generally only found inside stars, matter turns into a state called a **plasma**.
4) A lot of **energy** is released during nuclear fusion because the new, heavier nucleus has a **much higher binding energy per nucleon** (and so a lower total mass, see p.78). The energy released helps to **maintain the high temperatures** needed for further fusion reactions.
5) Although the energy released per reaction is generally **lower** in nuclear fusion than fission, the nuclei used in fusion have a **lower mass**, so a mole of the reactants in a fusion reaction weighs less than a mole of the reactants in a fission reaction (p.80). Gram for gram, fusion can release **more energy** than fission.
6) Scientists are trying to develop fusion reactors so that we can generate nuclear **electricity** without the waste you get from fission reactors, but they haven't yet succeeded in creating one that makes more electricity than it uses.

Example: In the Sun, **hydrogen nuclei** fuse to form **helium:**

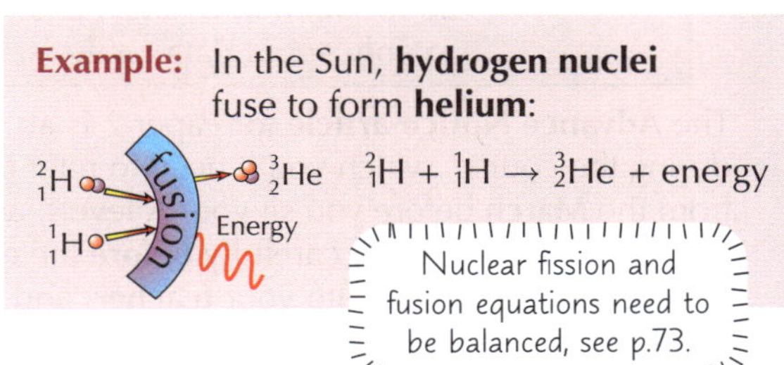

$^2_1H + ^1_1H \rightarrow ^3_2He + \text{energy}$

Nuclear fission and fusion equations need to be balanced, see p.73.

There's more on moles on p.38.

Practice Questions

Q1 What is meant by the term 'induced fission'?

Q2 Explain what is meant by the expressions 'chain reaction', 'fuel rods' and 'moderator' in terms of nuclear fission.

Q3 What are the similarities and differences between nuclear fusion and fission?

Exam Questions

Q1 Nuclear reactors use carefully controlled chain reactions to produce energy.

a) Describe and explain one feature of a nuclear reactor whose role is to control the rate of fission. Include an example of a suitable material for the feature you have chosen. [3 marks]

b) Explain what happens in a nuclear reactor during an emergency shut-down. [2 marks]

Q2 Discuss two advantages and two disadvantages of using nuclear fission to produce electricity. [4 marks]

If anyone asks, I've gone fission... that joke never gets old...

So, controlled nuclear fission reactions can provide a shedload of energy to generate electricity without producing pesky carbon dioxide, but nuclear energy has costs and risks too... Nothing's ever simple, is it?

Exam Structure and Technique

Good exam technique can make a big difference to your mark, so make sure you read this stuff carefully.

*Get Familiar With the **Exam Structure***

For A-level Physics, you'll be sitting **three papers**. You'll need to all the physics you've covered in both years of your course for **all three** exams.

You'll also do a Practical Endorsement as part of your A-level. It'll involve doing practicals throughout the course, and will be reported separately from your exam results.

Paper	Weighting
Paper 1 — Fundamentals of physics **2 hours 15 minutes** **110** marks: **30** for **Section A** (**multiple choice** questions) ≈ **20** for **Section B** (**short answer** questions) ≈ **60** for **Section C** (**short answer** and **extended response** questions)	**41%** of your A-level
Paper 2 — Scientific literacy in physics **2 hours 15 minutes** **100** marks: ≈ **30** for **Section A** (**short answer** questions) ≈ **45** for **Section B** (**short answer** and **extended response** questions) ≈ **25** for **Section C** (**short answer** and **extended response** questions, based on the **Advance Notice article** that you'll be given before the exam)	**37%** of your A-level
Paper 3 — Practical skills in physics **1 hour 30 minutes** **60 marks:** ≈ **40** for **Section A** (**short answer** and **extended response** questions) ≈ **20** for **Section B** (**short answer** and **extended response** questions, mainly about data analysis).	**22%** of your A-level

The **Advance Notice article** for Paper 2 is an article (or articles) on a topic related to something you have **studied** during the course, which you'll need to refer to when answering the questions in **Section C**. It'll be available from the **March** before you sit your A levels, and your teacher should give it to you at least 4 weeks before your exams. **Read it through** carefully **before** the exam — there won't be time to read it in the test. You'll probably go through and discuss it with your **teacher**, and you may want to do some **further research** on the subject it covers. You **can't take any notes** on it into the exam with you — you'll be given a **fresh copy** with your test paper.

*Make Sure you Think About **Exam Technique***

1) It's really important you read each question **carefully**, and give an answer that matches what you've been asked.
2) Look for **command words** in the question — they'll give you an idea of the **kind of answer** you should write. Commonly used command words for written questions are **state**, **describe**, **discuss** and **explain**:
    - If a question asks you to **state** something, you just need to give a **definition**, **example** or **fact**.
    - If you're asked to **describe** what happens in a particular situation, don't waste time explaining why it happens — that's not what the question is after.
    - For **discuss** questions, you'll need to include more **detail** — depending on the question you might need to cover what happens, what the effects are, and perhaps include a brief explanation of why it happens.
    - If a question asks you to **explain** why something happens you must give **reasons**, not just a description.
3) Look at **how many marks** a question is worth before answering. It'll tell you roughly **how much information** you need to include. See the next page for more about **wordy questions**.
4) The **number of marks** also tells you roughly **how long** to spend on a question — there's just over a minute per mark in the exam. If you get stuck on a question for too long, it may be best to **move on**.
5) The **multiple choice questions** are only worth **one mark each**, so it's not worth stressing over one for ages if you get stuck — you can always come back to it later if there's time at the end.
6) You don't have to work through the paper **in order** — you could leave questions on topics you find harder until last.

You may get some **weird questions** that seem to have nothing to do with anything you've learnt. **DON'T PANIC.** Every question will be something you can answer **using physics you know**, it just may be in a new **context**.

1) Answering these **trickier questions** will get you **top marks**, but make sure you cover the **easier marks** first.
2) All of the A-level exams could **pull together** ideas from different parts of physics, so check the question for any **keywords** that you recognise. For example, if a question talks about acceleration, think about the rules and equations you know, and whether any of them apply to the situation in the question.

Exam Structure and Technique

Be **Prepared** for **Practical Questions...**

1) **Practical skills** are a big part of your course — they're particularly important for **Paper 3**, but could come up in **Paper 1** and **Paper 2** as well.
2) You may have to **describe an experiment** to investigate something, or **answer questions** on an experiment you've been given. These could be experiments you've **met before**, or they could be **entirely new** to you.
3) All the questions will be based on physics that you've **covered**, but may include bits from different topics put together in ways you haven't seen before. Don't let this put you off, just **think carefully** about what's going on.
4) You need to be able to describe the results of experiments using **scientific language** (like **precision**, **accuracy** and **validity**). Make sure you're comfortable with **experimental uncertainties** (the ± bit you'll sometimes see written after a result), and that you can identify sources of **random** and **systematic error** that may have interfered with the results of an experiment.
5) You need to be able to **calculate errors** and **plot** and **interpret graphs** too.

...and **Wordy Ones**

For some questions, you'll need to write a slightly longer answer, where the '**quality** of your **extended response**' will be taken into account. You'll need to make sure you can develop a **clear** and **logical**, **well-structured line of reasoning**, backed up with **relevant information**.
You can avoid losing marks in these questions by making sure you do the following things:

1) Think about your answer before you write it. Your answer needs to be **logically structured** to get the top marks.
2) Make sure your answer is **relevant** to the question being asked and that you **explain** your ideas or argument **clearly**. It's dead easy to go off on a tangent.
3) Back up your points with **evidence** or **explanation**. You'll lose marks if you just make statements without supporting them.
4) Write in **whole sentences** and keep an eye on your **spelling**, **grammar** and **punctuation**. It'll help make sure your answer is clear and easy to read.

Questions like this will be marked in some way in the exam — e.g. with an asterisk (*). Check the instructions on the front of your paper to find out.

Example: A large group of people walk across a footbridge. When the frequency of the group's footsteps is 1 Hz, the bridge noticeably oscillates.
Describe the phenomenon causing the bridge to oscillate, and suggest what engineers could do to solve this problem. *[6 marks]*

Good Answer

The pedestrians provide a driving force on the bridge, causing it to oscillate. At around 1 Hz, the driving frequency from the pedestrians is roughly equal to the natural frequency of the bridge, causing it to resonate. The amplitude of the bridge's oscillations when resonating at 1 Hz will be greater than at any other driving frequency. The oscillations at this frequency are large enough to be noticed by pedestrians. Engineers could fix this problem by critically damping the bridge to stop any oscillations as quickly as possible. They could also adjust the natural frequency of the bridge so that it was not so close to a known walking frequency of large groups of people.

Bad Answer

resonance
driving frequency = natural frequency
damping

There's nothing fundamentally wrong with the physics in the bad answer, but you'd miss out on some nice easy marks just for not bothering to link the physics with the context given and not putting your answer into proper sentences.

The penultimate joke in the book better be good... here goes... Oh, I've run out of space...

Make sure you use your exam timetable to plan your revision carefully, and give yourself enough time to revise everything. If you leave it all till the night before, you won't be at your best on the day and might not do as well in the exam...

Maths Skills

At least 40% of the marks up for grabs in A-level Physics will require maths skills, so make sure you know your stuff. As well as being given some tricky calculations, you could be asked to work with exponentials and logarithms and work out values from log graphs. And it's easy when you know how...

Be Careful With Calculations

1) In calculation questions you should always **show your working** — you may get some marks for your **method** even if you get the answer wrong.
2) Don't **round** your answer until the **very end**. A lot of calculations in A-level Physics are quite **long**, and if you round too early you could introduce errors to your final answer.
3) Be careful with **units**. Lots of formulas require quantities to be in specific units (e.g. time in seconds), so it's best to **convert** any numbers you're given into these before you start. And obviously, if the question **tells** you which units to give your **answer** in, don't throw away marks by giving it in different ones.
4) You should give your final answer to the same number of **significant figures** as the data that you use from the question with the **least number** of significant figures (or one more). If you can, write out the **unrounded answer**, then your **rounded** answer with the number of significant figures you've given it to — it shows you know your stuff.

Many Relationships in Physics are Exponential

A fair few of the relationships you need to know about in A-level Physics are **exponential** — where the **rate of change** of a quantity is **proportional** to the **amount** of the quantity left. Here are a few that crop up in the A-level course (if they don't ring a bell, go have a quick read about them)...

The equation for finding the Boltzmann factor also includes an exponential function (p.44).

Charge on a capacitor — the decay of charge on a discharging capacitor is proportional to the amount of charge left on the capacitor: $Q = Q_0 e^{\frac{-t}{CR}}$ (see p.9)

There are also exponential relationships for I and V and for charging capacitors.

Radioactive decay — the rate of decay of a radioactive sample is proportional to the **number of undecayed nuclei** in the sample: $N = N_0 e^{-\lambda t}$ (see p.6)

The activity of a radioactive sample behaves in the same way.

You can Plot Exponential Relations Using the Natural Log, ln

1) Say you've got two variables, x and y, which are related to each other by the formula $y = ke^{-ax}$ (where k and a are constants).
2) The **natural logarithm** of x, **ln** x, is the power to which e (the base) must be raised to to give x.
3) So, by definition, $e^{\ln x} = x$ and $\ln(e^x) = x$.
So far so good... now you need some **log rules**:

A logarithm can be to any base you want. Another common one is 'base 10' which is usually written as '$\log_{10}$' or just 'log'.

$\ln(AB) = \ln A + \ln B$ $\qquad \ln\left(\frac{A}{B}\right) = \ln A - \ln B$ $\qquad \ln x^n = n \ln x$

These log rules work for all logs (including the natural logarithm) and you're given them on your formula sheet.

4) So, for $y = ke^{-ax}$, if you take the natural log of both sides of the equation you get:

$$\ln y = \ln(ke^{-ax}) = \ln k + \ln(e^{-ax}) \longrightarrow \ln y = \ln k - ax$$

5) Then all you need to do is plot (ln y) against x, and Eric's your aunty:

You get a **straight-line** graph with (**ln** k) as the **vertical intercept**, and **$-a$** as the **gradient**.

Maths Skills

You Might be Asked to find the **Gradient** of a Log Graph

This log business isn't too bad when you get your head around which bit of the log graph means what.

Example: The graph shows the radioactive decay of isotope X.

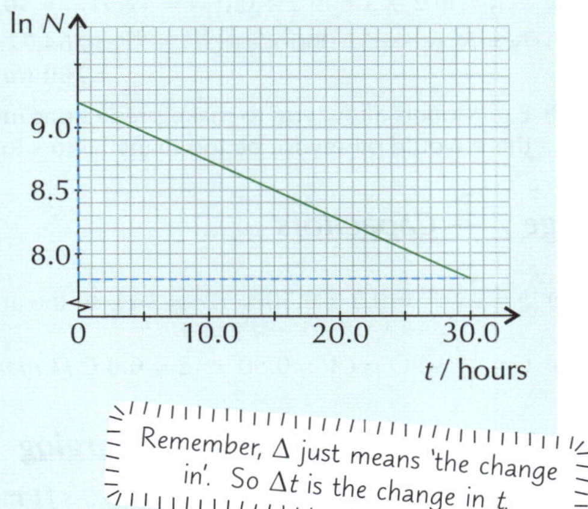

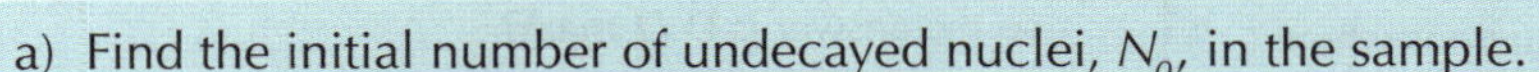

a) Find the initial number of undecayed nuclei, N_0, in the sample.

You know that the number of undecayed nuclei in a sample, N, is related to the initial number of undecayed nuclei, N_0, by the equation $\boldsymbol{N = N_0 e^{-\lambda t}}$.

So: $\ln N = \ln N_0 - \lambda t$

The y-intercept of the graph is $\ln N_0 = 9.2$

$N_0 = e^{9.2} = 9897.129... =$ **9900 nuclei (to 2 s.f.)**

Remember, Δ just means 'the change in'. So Δt is the change in t.

b) Find the decay constant λ of isotope X.

$-\lambda$ is the gradient of the graph, so: $\lambda = \frac{\Delta \ln N}{\Delta t} = \frac{9.2 - 7.8}{30.0 \times 60 \times 60} = \mathbf{1.3 \times 10^{-5}\ s^{-1}}$ **(to 2 s.f.)**

You can Plot **Any Power Law** as a **Log-Log Graph**

You can use logs to plot a straight-line graph of **any power law** — it doesn't have to be an exponential.

Say the relationship between two variables x and y is:

$$y = kx^n$$

Take the **log** (base 10) of both sides to get:

$$\log y = \log k + n \log x$$

So **log *k*** will be the ***y*-intercept** and ***n*** will be the **gradient** of the graph.

When it came to logs, Geoff always took time to smell the flowers...

Example: A physicist carries out an experiment to determine the nuclear radius R (in m) of various elements, which have nucleon number A (number of protons + number of neutrons). She plots a line of best fit for her results on a graph of $\log R$ against $\log A$. Part of the graph is shown. Using the equation $R = R_0 A^{1/3}$, find the value of the constant R_0 from the graph.

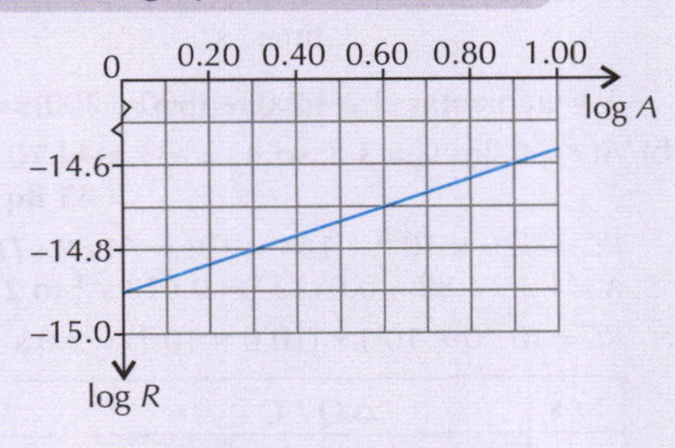

First take logs of both sides:

$$\log R = \log (R_0 A^{1/3}) = \log R_0 + \log A^{1/3}$$
$$= \log R_0 + \frac{1}{3} \log A$$

Comparing this to the equation of a straight line (in the form $y = mx + c$), you can see that the gradient of the graph is $\frac{1}{3}$ and the vertical intercept is $\log R_0$.

So, reading from the graph, the vertical intercept is about –14.9.

$\log R_0 = -14.9$, so $R_0 = 10^{-14.9} = 1.258... \times 10^{-15} =$ **1.3 fm (to 2 s.f.)**

If $a = \log_{10} b$, then $b = 10^a$.

Lumberjacks are great musicians — they have a natural logarhythm...

Well, that's it folks. Crack open the chocolate bar of victory and know you've earned it. Only the tiny detail of the actual exam to go... ahem. Make sure you know which bit means what on a log graph and you'll pick up some nice easy marks. Other than that, stay calm, be as clear as you can and good luck — I've got my fingers, toes and eyes crossed for you.

Answers

Module 5: Section 1 — Creating Models

Page 6 — Radioactivity and Exponential Decay

1 a) $A = \lambda N = 0.014 \times 51\ 000 = 714 =$ **710 Bq (to 2 s.f.)** *[1 mark]*

b) $T_{1/2} = \ln 2 \div \lambda = \ln 2 \div 0.014 = 49.51... =$ **50 s (to 2 s.f.)** *[1 mark]*

c) $N = N_0 e^{-\lambda t} = 51\ 000 \times e^{-(0.014 \times 300)} = 764.77...$
$=$ **760 (to 2 s.f.)** *[1 mark]*

d) E.g. Models allow you to make predictions in situations where there would otherwise be too many factors to consider *[1 mark]*.

Page 7 — Capacitors

1 a) $E = \frac{1}{2}CV^2 = 0.5 \times 0.50 \times 12^2 = 36$ J, so the answer is **A** *[1 mark]*

b) $C = \frac{Q}{V}$ so $Q = CV = 0.50 \times 12 =$ **6.0 C** *[1 mark]*

Page 10 — Charging and Discharging

1 a) $RC = (1.0 \times 10^3) \times (250 \times 10^{-6}) = 0.25$ s *[1 mark]*

$V = V_0(1 - e^{\frac{-t}{RC}}) = 6.0 \times (1 - e^{\frac{-0.25}{0.25}}) = 3.792...$
$=$ **3.8 V (to 2 s.f.)** *[1 mark]*

b) $Q = Q_0 e^{\frac{-t}{RC}}$ so after 0.7 seconds: $Q = Q_0 e^{\frac{-0.7}{0.25}}$
$= Q_0 \times 0.06$ (to 1 s.f.) *[1 mark]*
So there is **6% (to 1 s.f.)** of the initial charge left on the capacitor after 0.7 seconds *[1 mark]*.

2 a) The charge falls to 37% after RC seconds *[1 mark]*, so $t = (1.6 \times 10^3) \times (320 \times 10^{-6}) =$ **0.512 seconds** *[1 mark]*

b) $\frac{dQ}{dt} = -\frac{Q}{RC} = -\frac{5.5 \times 10^{-3}}{0.512}$
$= -0.0107... =$ **-0.011 Cs^{-1} (to 2 s.f.)** *[1 mark]*

Page 13 — Modelling Decay

1 a)

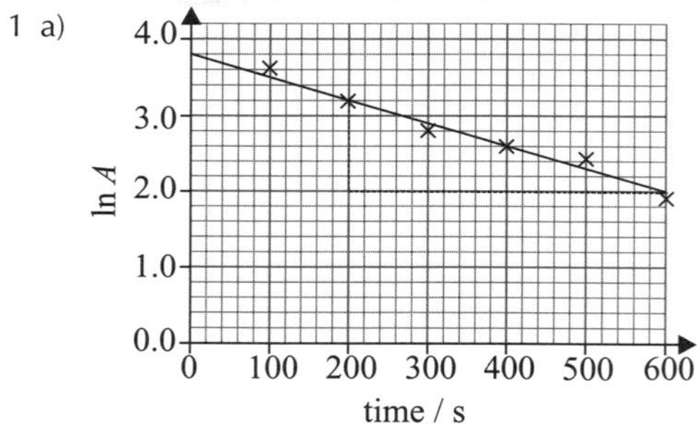

λ = gradient $= (3.2 - 2.0) \div (600 - 200) =$ **3.0×10^{-3} s^{-1}** *[1 mark]*

b) At $t = 0$, $\ln A_0 = 3.8$, so $A_0 = e^{3.8} = 44.70...$
$=$ **45 Bq (to 2 s.f.)** *[1 mark]*

2 $RC = 520 \times 10^{-6} \times 144 \times 10^3 = 74.88$ s *[1 mark]*
$\lambda = 1 \div 74.88 = 0.0133... =$ **0.013 s^{-1} to 2 s.f.** *[1 mark]*

3 $RC = (0.20 \times 10^6) \times (10.0 \times 10^{-6}) = 2.0$ s

t / s	ΔQ / C	Q / C
0		0.050
0.50	$= -(0.05 \div 2.0) \times 0.50$ $= -0.0125$	$= 0.050 + (-0.0125)$ $= 0.0375$
1.0	$= -(0.0375 \div 2.0) \times 0.50$ $= -9.375 \times 10^{-3}$	$= 0.0375 + (-9.375 \times 10^{-3})$ $= 0.028125$

So the answer is **0.028 C (to 2 s.f.)**
[1 mark for each correct iteration]

Page 15 — Simple Harmonic Motion

1 a) Simple harmonic motion is an oscillation in which the restoring force is always directly proportional to the object's displacement *[1 mark]* and always acts towards the midpoint *[1 mark]*.

b) The acceleration of a falling bouncy ball is due to gravity. This acceleration is constant, so the motion is not SHM. *[1 mark]*.

2 a) Maximum velocity $= \omega A = 2\pi f A = 2\pi \times 1.5 \times 0.050 = 0.47$ ms^{-1} so the answer is **D** *[1 mark]*

b) Clock started when object released, so $x = A\cos(2\pi ft)$ *[1 mark]*
$x = 0.050 \times \cos(2\pi \times 1.5 \times 0.10) = 0.02938...$
$=$ **0.029 m (to 2 s.f.)** *[1 mark]*

Page 17 — Investigating Simple Harmonic Motion

1 a) Extension of spring $= 0.200 - 0.100 = 0.100$ m *[1 mark]*.
$F = kx$ and $F = mg$ so $k = mg/x = 0.10 \times 9.81 / 0.100 = 9.81$
$=$ **9.8 Nm^{-1} (to 2 s.f.)** *[1 mark]*

b) i) maximum extension $= 0.10 + 0.020 = 0.12$ m
$E = \frac{1}{2}kx^2 = \frac{1}{2} \times 9.81 \times 0.12^2 = 0.070632$
$=$ **0.071 J (to 2 s.f.)** *[1 mark]*

ii) $T = 2\pi\sqrt{\frac{m}{k}}$ so $T \propto \sqrt{m}$ and $m \propto T^2$ *[1 mark]*
So the mass needed is $2^2 \times 0.10 =$ **0.40 kg** *[1 mark]*

Alternatively, you could calculate the new period of the oscillations (2×0.63), then rearrange the equation $T = 2\pi\sqrt{\frac{m}{k}}$ to find m.

2 $5T_{\text{short pendulum}} = 3T_{\text{long pendulum}}$
$T = 2\pi\sqrt{\frac{L}{g}}$ so: $5 \times 2\pi\sqrt{\frac{0.20}{g}} = 3 \times 2\pi\sqrt{\frac{L}{g}}$ *[1 mark]*
multiplying both sides by $\frac{\sqrt{g}}{2\pi}$ gives: $5 \times \sqrt{0.20} = 3 \times \sqrt{L}$
Squaring both sides gives: $25 \times 0.20 = 9L$ *[1 mark]*
so $L = 25 \times 0.20 \div 9 = 0.555... =$ **0.56 m (to 2 s.f.)** *[1 mark]*

Page 19 — Modelling Simple Harmonic Motion

1 a) $k \div m = 0.82 \div 0.025 = 32.8$ Nm^{-1}kg

t / s	Δv / ms^{-1}	v ms^{-1}
0		0
0.05	$= -32.8 \times 0.024 \times 0.05$ $= -0.03936$	$= 0 + (-0.03936)$ $= -0.03936$
0.10	$= -0.0361...$	$= -0.0754...$
0.15	$= -0.0299...$	$= -0.105...$

...

Δx / m	x / m
	0.024
$= -0.03936 \times 0.05$ $= -0.001968$	$= 0.024 + (-0.001968)$ $= 0.022032$
$= -0.00377...$	$= 0.0182...$
$= -0.00527...$	$= 0.0129...$

So at $t = 0.15$, $x =$ **0.013 m (to 2 s.f.)** *[1 mark for each correct stage of iteration, to a total of 3 marks. Otherwise 1 mark for an attempt at an iterative method.]*

b) $d^2x/dt^2 = -\frac{k}{m}x$
d^2x/dt^2 is at its maximum when $x = A$ *[1 mark]*
The bob was pulled 0.024 m from its equilibrium position, so $A = 0.024$ m
so $d^2x/dt^2 = -\frac{0.82}{0.025} \times 0.024 = -0.7872$
$=$ **-0.79 ms^{-2} (to 2 s.f.)** *[1 mark]*

You'd also get the mark if you left out the negative sign — a positive value of d^2x/dt^2 just means the pendulum is on the other side of its swing.

Answers

Page 21 — Free and Forced Vibrations

1 a) E.g.

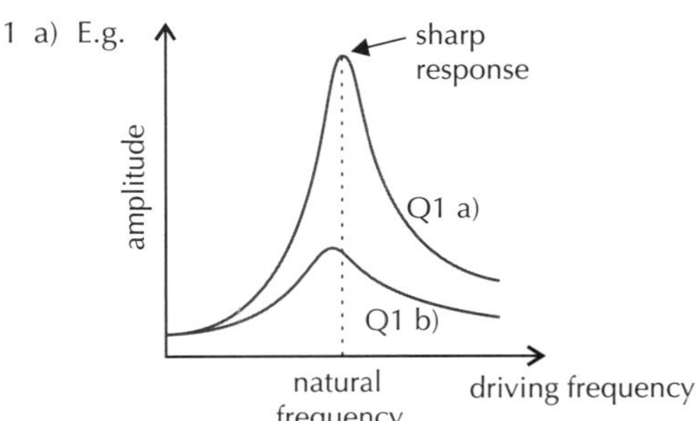

[1 mark for showing a peak at the natural frequency, 1 mark for a sharp peak.]

b) ***[See graph in part a). 1 mark for a smaller peak at the natural frequency.]***
The peak will actually be slightly to the left of the natural frequency due to the damping, but you'll get the mark if the peak is at the same frequency in the diagram.

2 A system is critically damped if it returns to rest in the shortest time possible when it's displaced from equilibrium and released ***[1 mark]***. It is used in e.g. the suspension in a car ***[1 mark]***.

Module 5: Section 2 — Out Into Space

Page 23 — Circular Motion

1 a) $v = r\omega = (1.5 \times 10^{11}) \times (2.0 \times 10^{-7})$ ***[1 mark]***
$= 30\ 000\ \text{ms}^{-1} =$ **30 kms^{-1}** ***[1 mark]***

b) i) $F = m\omega^2 r = (6.0 \times 10^{24}) \times (2.0 \times 10^{-7})^2 \times (1.5 \times 10^{11})$ ***[1 mark]***
$=$ **3.6 × 10^{22} N** ***[1 mark]***

ii) The gravitational force between the Sun and the Earth ***[1 mark]***.

2 a) $a = v^2 \div r = 15^2 \div 12 = 18.75 =$ **19 ms^{-2} (to 2 s.f.)** ***[1 mark]***

b) Since $a = \omega^2 r$, $\omega^2 = \frac{a}{r} = \frac{18.75}{12}$, so $\omega = 1.25$ rad s^{-1} ***[1 mark]***

$\omega = 2\pi \div T$, so $T = 2\pi \div \omega = 5.026... =$ **5.0 s (to 2 s.f.)** ***[1 mark]***

Page 25 — Gravitational Fields

1 a) $F_{grav} = -\frac{GmM}{r^2} = -\frac{(6.67 \times 10^{-11}) \times 25 \times (7.35 \times 10^{22})}{(1740 \times 10^3)^2}$ ***[1 mark]***

$= -40.48... =$ **–40 N (to 2 s.f.)** ***[1 mark]***

b) i) $\frac{1}{2}mv^2 = mgh$ ***[1 mark]***,
so $v = \sqrt{2gh} = \sqrt{2 \times 1.62 \times 4.5}$
$= 3.818... =$ **3.8 ms^{-1} (to 2 s.f.)** ***[1 mark]***

ii) As the rock rises in the gravitational field, its kinetic energy is transferred into gravitational potential energy and it slows down ***[1 mark]***. It stops gaining height once all of its initial kinetic energy has been transferred into gravitational potential energy ($\frac{1}{2}mv^2 = mgh$) ***[1 mark]***. It then changes direction and begins to fall, losing gravitational potential energy and gaining kinetic energy until it collides with the surface of the Moon ***[1 mark]***.

Page 27 — Gravitational Potential and Orbits

1 a) $r = 6400 + 200 = 6600$ km

$g = -\frac{GM}{r^2} = -\frac{(6.67 \times 10^{-11}) \times (5.98 \times 10^{24})}{(6600 \times 10^3)^2}$ ***[1 mark]***

$= -9.156... =$ **–9.2 Nkg^{-1} (to 2 s.f.)** ***[1 mark]***

b) $V_{grav} = -\frac{GM}{r} = g \times r$
$= -9.156... \times 6600 \times 10^3$ ***[1 mark]*** $= -6.04... \times 10^7$
$=$ **–6.0 × 10^7 Jkg^{-1} (to 2 s.f.)** ***[1 mark]***

c) $E_{grav} = -\frac{GmM}{r} = V_{grav} \times m = -6.04... \times 10^7 \times 3015$
$= -1.822... \times 10^{11} =$ **–1.8 × 10^{11} J (to 2 s.f.)** ***[1 mark]***

d) $v = \sqrt{\frac{GM}{r}} = \sqrt{\frac{6.67 \times 10^{-11} \times 5.98 \times 10^{24}}{6600 \times 10^3}}$ ***[1 mark]***
$= 7773.9... =$ **7800 ms^{-1} (to 2 s.f.)** ***[1 mark]***

Module 5: Section 3 — Our Place in the Universe

Page 29 — Using Astronomical Scales

1 a) $m = -2.5 \log_{10} I + \text{constant}$, so $\log_{10} I = \frac{m - \text{constant}}{-2.5}$

So $I = 10^{\frac{m - \text{constant}}{-2.5}}$ ***[1 mark]***

So $\frac{I_1}{I_2} = \frac{10^{\frac{m_1 - \text{constant}}{-2.5}}}{10^{\frac{m_2 - \text{constant}}{-2.5}}}$ ***[1 mark]***

$= 10^{\left(\frac{m_1 - \text{constant}}{-2.5} - \frac{m_2 - \text{constant}}{-2.5}\right)}$

$= 10^{\frac{1}{-2.5}(m_1 - \text{constant} - m_2 + \text{constant})}$

$= 10^{\frac{1}{-2.5}(m_1 - m_2)}$

$= 10^{\frac{1}{2.5}(m_2 - m_1)} = 10^{\frac{1}{2.5}(2 - 1)} = 10^{\frac{1}{2.5}} =$ **10$^{\frac{2}{5}}$** ***[1 mark]***

Remember your power laws: $x^a \div x^b = x^{a-b}$.

b) From part a) $\frac{I_1}{I_2} = 10^{\frac{1}{2.5}(m_2 - m_1)}$

$m_1 = 1$, $m_2 = 6$, so $\frac{I_1}{I_2} = 10^{\frac{1}{2.5}(6 - 1)} = 10^2$
$=$ **100 times greater intensity** ***[1 mark]***

2 a) A light-year is the distance travelled by electromagnetic waves through a vacuum in one year ***[1 mark]***.

b) Average number of seconds in a year $= 365.25 \times 24 \times 60 \times 60$
$= 3.15... \times 10^7$ s

Distance = speed × time $= 3.00 \times 10^8 \times 3.15... \times 10^7$ ***[1 mark]***
$= 9.467... \times 10^{15}$
$=$ **9.47 × 10^{15} m (to 3 s.f.)** ***[1 mark]***

You'd also get the marks if you used 3.16×10^7 s for the number of seconds in a year, it's given in your data and formulae booklet.

c) To see something, light from the object must reach us. Light travels at a finite speed, so it takes time for that to happen ***[1 mark]***. The universe has a finite age, so we can only see so far ***[1 mark]***.

Page 31 — Astronomical Distances and Velocities

1 a) $2d = ct$ ***[1 mark]***, so $d = [(3.00 \times 10^8) \times (4.6 \times 60)] \div 2$ ***[1 mark]***
$d = (8.28 \times 10^{10}) \div 2 =$ **4.14 × 10^{10} m (to 3 s.f.)** ***[1 mark]***

b) Any two of: The speed of the radio waves (the speed of light) is constant ***[1 mark]***. / The time taken for the radio waves to reach Venus is the same as the time taken for them to return to Earth ***[1 mark]***. / The speed of Venus relative to the Earth is much less than the speed of light ***[1 mark]***.

2 a) First pulse: $d = ct \div 2 = (3.00 \times 10^8 \times 2.8) \div 2$
$= 4.2 \times 10^8$ m ***[1 mark]***

Second pulse: $d = ct \div 2 = (3.00 \times 10^8 \times 3.8) \div 2$
$= 5.7 \times 10^8$ m ***[1 mark]***

Change in distance $= (5.7 \times 10^8) - (4.2 \times 10^8) = 1.5 \times 10^8$ m
So relative velocity $v = d \div t = (1.5 \times 10^8) \div 1$
$=$ **1.5 × 10^8 ms^{-1}** ***[1 mark]***

b) $t = \frac{t_0}{\sqrt{1 - \frac{v^2}{c^2}}} = \frac{1}{\sqrt{1 - \frac{(1.5 \times 10^8)^2}{(3.00 \times 10^8)^2}}}$ ***[1 mark]***

$= 1.154... =$ **1.2 s (to 2 s.f.)** ***[1 mark]***

Answers

Page 33 — The Hot Big Bang Theory

1 a) Hubble's law suggests that the universe originated with the hot big bang ***[1 mark]*** and has been expanding ever since ***[1 mark]***.

b) i) $H_0 = v / d = 75$ kms^{-1} / 1 Mpc
75 kms^{-1} = 75 × 10^3 ms^{-1} and 1 Mpc = 3.09 × 10^{22} m
So H_0 = (75 × 10^3 ms^{-1}) ÷ (3.09 × 10^{22} m)
= 2.427... × 10^{-18} s^{-1} = **2.43 × 10^{-18} s^{-1} (to 3 s.f.)**
[1 mark for the correct value, 1 mark for the correct unit]

ii) $t = 1/H_0$ so t = 1 ÷ (2.427... × 10^{-18}) = 4.12 × 10^{17} s ***[1 mark]***
(4.12 × 10^{17}) ÷ (3.16 × 10^7) = 1.303... × 10^{10}
= **13 billion years (to 2 s.f.)** ***[1 mark]***
So the observable universe has a radius of
13 billion light-years (to 2 s.f.) ***[1 mark]***.

2 a) $\frac{\Delta\lambda}{\lambda} = \frac{v}{c}$ so $v = c\frac{\Delta\lambda}{\lambda}$ = 3.00 × 10^8 × $\frac{(890\times10^{-9})-(650\times10^{-9})}{650\times10^{-9}}$
= 1.107... × 10^8
= **1.1 × 10^8 ms^{-1} (to 2 s.f.)** ***[1 mark]***
away from us ***[1 mark]***
You can tell the galaxy is moving away from us as the wavelength of the spectral line has increased (it has been red shifted).

b) $v = H_0 d$ so $d = v / H_0$ = (1.107... × 10^8) ÷ (2.4 × 10^{-18})
= 4.615... × 10^{25} m ***[1 mark]***
(4.615... × 10^{25}) ÷ (9.5 × 10^{15})
= 4.858... ×10^9 = **4.9 × 10^9 ly (to 2 s.f.)** ***[1 mark]***

c) $\frac{\Delta\lambda}{\lambda} = \frac{v}{c}$ is only valid if $v << c$ — it isn't in this case ***[1 mark]***.

3 It suggests that the very early universe was very hot, producing lots of electromagnetic radiation ***[1 mark]*** and that the universe's expansion has stretched the radiation into the microwave region ***[1 mark]***.

Module 5: Section 4 — Matter: Very Simple

Page 35 — Temperature and Specific Thermal Capacity

1 a) There is a net transfer of thermal energy from the water to the stone, as the stone is colder than the water. The water loses energy, so its temperature decreases. ***[1 mark]***.

b) $\Delta E = mc\Delta\theta$ = 0.025 × 4180 × (12 – 15)
= –313.5 = **–310 J (to 2 s.f.)** ***[1 mark]***

2 a) Energy needed = $\Delta E = mc\Delta\theta$ = 0.93 × 210 × (34.2 – 18.5)
= 3066.21 J ***[1 mark]***
$E = W = VIt$ so $t = E \div VI$ = 3066.21 ÷ (32 × 5.0)
= 19.163... = **19 s (to 2 s.f.)** ***[1 mark]***

b) The actual time taken will be longer ***[1 mark]*** because energy is lost heating the insulating container ***[1 mark]*** / because energy is lost due to resistance in the circuit ***[1 mark]***.

Page 38 — Ideal Gases

1 At constant volume, the pressure of a gas is directly proportional to its absolute temperature ***[1 mark]***.
The graph plotted will be a straight line ***[1 mark]***. It will cross the temperature axis at absolute zero (–273.15 °C or 0 K) ***[1 mark]***.

2 a) number of moles, n = mass ÷ molar mass = 0.014 ÷ 0.028
= 0.50 ***[1 mark]***
number of particles, $N = nA_v$ = 0.50 × 6.02 × 10^{23} = 3.01 × 10^{23}
= **3.0 × 10^{23} (to 2 s.f.)** ***[1 mark]***

b) T = 27°C = 27 + 273 = 300 K ***[1 mark]***
$pV = NkT$ so $p = NkT \div V$
= ((3.01 × 10^{23}) × (1.38 × 10^{-23}) × 300) ÷ 0.010
= 124 614 = **120 000 Pa (to 2 s.f.)** ***[1 mark]***

c) The pressure would also halve because it is proportional to the number of molecules — $pV = NkT$ ***[1 mark]***.

3 $pV = NkT$ so $pV \div T = Nk$ = constant ***[1 mark]***
$(pV \div T)_{ground} = (pV \div T)_{raised}$
$\frac{(1.0\times10^5)\times10.0}{293} = \frac{p_{raised}\times25}{260}$
$p_{raised} = \frac{(1.0\times10^5)\times10.0\times260}{293\times25}$ = 35 494.88...
= **35 000 Pa (to 2 s.f.)** ***[1 mark]***

Page 41 — Kinetic Theory and Internal Energy

1 a) $pV = \frac{1}{3}Nm\overline{c^2}$ so $\overline{c^2} = \frac{3pV}{Nm} = \frac{3\times1.0\times10^5\times7.0\times10^{-5}}{2.0\times10^{22}\times6.6\times10^{-27}}$
= 159 090.90... ***[1 mark]***
$\sqrt{\overline{c^2}} = \sqrt{159\,090.90...}$ = 398.86... = **400 ms^{-1} (to 2 s.f.)** ***[1 mark]***

b) The increase in temperature will increase the average speed and therefore momentum of the gas particles ***[1 mark]***. This means that as the particles hit the walls of the flask there will be a greater change in momentum, meaning the impulse of the collisions is greater, and thus exerting a greater force on the walls of the flask ***[1 mark]***. As the particles are travelling faster they will also hit the walls (and exert a force) more frequently ***[1 mark]***. This leads to an increase in pressure (as pressure = force ÷ area) ***[1 mark]***.

2 a) speed = distance / time so time = distance / speed = 8.0 ÷ 410
= 0.0195...
= **0.020 s (to 2 s.f.)** ***[1 mark]***

b) Although the particles are moving at an average of 410 ms^{-1}, they are frequently colliding with air molecules/other air freshener particles ***[1 mark]***. This means their motion follows a random walk, so their movement in any one direction is limited and they only slowly move from one end of the room to the other ***[1 mark]***.

c) average energy per particle = $\frac{3}{2}kT$
= $\frac{3}{2}$ × (1.38 × 10^{-23}) × (273 + 22)
= 6.1065 × 10^{-21} = **6.1 × 10^{-21} J (to 2 s.f.)** ***[1 mark]***
Remember, you need the temperature in kelvins, not degrees Celsius.

Module 5: Section 5 — Matter: Hot or Cold

Page 43 — Activation Energy

1 a) Average thermal energy is approximately kT.
i) kT = 1.38 × 10^{-23} × 290
= 4.002 × 10^{-21} = **4.0 × 10^{-21} J (to 2 s.f.)** ***[1 mark]***
ii) kT = 1.38 × 10^{-23} × 360
= 4.968 × 10^{-21} = **5.0 × 10^{-21} J (to 2 s.f.)** ***[1 mark]***

b) An activation energy is needed to break the strong attractive forces between the particles in the oil ***[1 mark]***.
The higher average thermal energy of the particles at 360 K means they are more likely to have an energy greater than the activation energy ***[1 mark]***, so more particles will be able to break the attractive forces and so the oil will be less viscous ***[1 mark]***.

Page 45 — The Boltzmann Factor

1 a) kT = 1.38 × 10^{-23} × 298 = 4.1124 × 10^{-21}
= **4.11 × 10^{-21} J (to 3 s.f.)** ***[1 mark]***

b) Proportion of molecules with energy of at least E_A is approximately given by the Boltzmann factor, $e^{-\frac{E_A}{kT}}$:
At 298 K: $\frac{E_A}{kT}$ = (9.4 × 10^{-20}) ÷ (4.1124 × 10^{-21})
= 22.857... ***[1 mark]***
So $e^{-\frac{E_A}{kT}} = e^{-22.857...}$
= 1.183... × 10^{-10} = **1.2 × 10^{-10} (to 2 s.f.)** ***[1 mark]***

Answers

c) Rate of reaction is approximately proportional to the Boltzmann factor, $e^{-\frac{E_A}{kT}}$:

From b), at 298 K, $e^{-\frac{E_A}{kT}} = 1.183... \times 10^{-10}$

At 313 K: $\frac{E_A}{kT} = (9.4 \times 10^{-20}) \div (1.38 \times 10^{-23} \times 313)$
$= (9.4 \times 10^{-20}) \div (4.3194 \times 10^{-21})$
$= 21.762...$ ***[1 mark]***

So at 313 K, $e^{-\frac{E_A}{kT}} = e^{-21.762...} = 3.538... \times 10^{-10}$ ***[1 mark]***

So the rate of reaction increases by a factor =
(rate of reaction at 313 K) ÷ (rate of reaction at 298 K)
$= (3.538... \times 10^{-10}) \div (1.183... \times 10^{-10})$
$= 2.990... =$ **3.0 (to 2 s.f.)** ***[1 mark]***

Module 6: Section 1 — Electromagnetism

Page 47 — Magnetic Fields

1 a) A current-carrying wire at right angles to an external uniform magnetic field feels a force because the field around the wire and the external field are added together. This can change the shape of the resultant flux lines ***[1 mark]***. Flux lines have a tendency to contract (get shorter) and straighten, which causes an electromagnetic force that pushes on the wire ***[1 mark]***.

b) The force acts horizontally towards the observer ***[1 mark]***.

2 a) $F = ILB = 3.0 \times 0.040 \times 2.0 \times 10^{-5}$ ***[1 mark]***
$= 2.4 \times 10^{-6}$ N ***[1 mark]***

b) The force is zero ***[1 mark]*** because there is no component of the current that is perpendicular to the magnetic field ***[1 mark]***.

Page 49 — Electromagnetic Induction

1 a) $\phi = BA = (2.0 \times 10^{-3}) \times 0.23 = 4.6 \times 10^{-4}$ Wb ***[1 mark]***
flux linkage $= \phi N = (4.6 \times 10^{-4}) \times 150 = 0.069$ Wb ***[1 mark]***

b) Change in flux linkage $= (3.5 \times 10^{-4} \times 150) - 0.069$
$= -0.0165$ Wb

$\varepsilon = -\frac{d(\phi N)}{dt} = -\frac{-0.0165}{2.5}$ ***[1 mark]***
$= 6.6 \times 10^{-3}$ V ***[1 mark]***

The e.m.f. is positive to oppose the reduction in magnetic flux that caused it ***[1 mark]***.

The positive e.m.f. induces a magnetic field around the coil that tries to 'top up' the falling magnetic field back to its initial value — it opposes the reduction in the uniform magnetic field.

Page 51 — Transformers and Dynamos

1 a) An ideal transformer is one that is 100% efficient. ***[1 mark]***

b) $\frac{V_1}{V_2} = \frac{N_1}{N_2}$ so, $N_2 = \frac{V_2}{V_1} \times N_1 = \frac{45}{9.0} \times 150$ ***[1 mark]***
$= 750$ turns ***[1 mark]***

c) $\frac{I_2}{I_1} = \frac{N_1}{N_2}$ so, $I_2 = \frac{N_1}{N_2} \times I_1 = \frac{150}{750} \times 1.5$ ***[1 mark]*** $= 0.30$ A ***[1 mark]***

Module 6: Section 2 — Charge and Field

Page 53 — Electric Fields

1 a) $E_{electric} = \frac{kQ}{r^2} = \frac{Q}{4\pi\varepsilon_0 r^2}$

$= \frac{1.60 \times 10^{-19}}{4\pi \times (8.85 \times 10^{-12}) \times (1.00 \times 10^{-10})^2}$ ***[1 mark]***

$= 1.4386... \times 10^{11}$

= **1.44 × 10¹¹ NC⁻¹ (or Vm⁻¹) (to 3 s.f.)** ***[1 mark]***

b) electric potential energy $= \frac{kQq}{r} = \frac{Qq}{4\pi\varepsilon_0 r}$

$= \frac{(1.60 \times 10^{-19}) \times (6.4 \times 10^{-19})}{4\pi \times (8.85 \times 10^{-12}) \times (3.25 \times 10^{-9})}$ ***[1 mark]***

$= 2.833... \times 10^{-19} =$ **2.8 × 10⁻¹⁹ J (to 2 s.f.)** ***[1 mark]***

c) 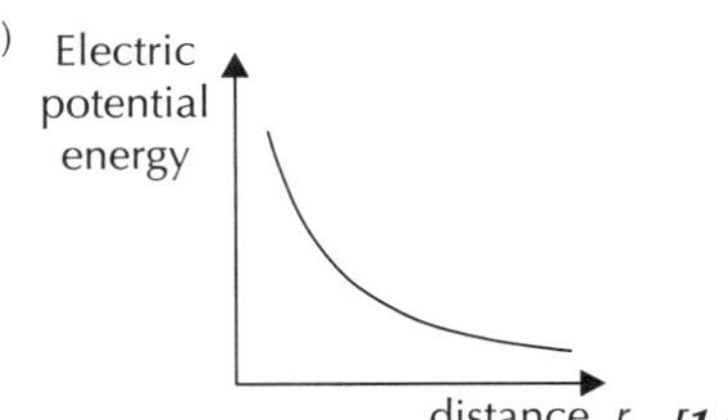

[1 mark]

d) The electric force $F_{electric}$ due to the electric field at that point ***[1 mark]***.

Page 55 — Electric Potential

1 $V_{electric} = \frac{kQ}{r} = \frac{Q}{4\pi\varepsilon_0 r}$

$= \frac{-1.60 \times 10^{-19}}{4\pi \times (8.85 \times 10^{-12}) \times 0.00100}$ ***[1 mark]***

$= -1.438... \times 10^{-6} =$ **−1.44 × 10⁻⁶ V (to 3 s.f.)** ***[1 mark]***

2 $V_{electric} = \frac{kQ}{r} = \frac{Q}{4\pi\varepsilon_0 r}$, so $Q = V_{electric} \times 4\pi\varepsilon_0 r$
$= 500.0 \times 4 \times \pi \times (8.85 \times 10^{-12}) \times (3.5 \times 10^{-3})$ ***[1 mark]***
$= 1.94... \times 10^{-10} =$ **1.95 × 10⁻¹⁰ C (to 3 s.f.)** ***[1 mark]***

3 a) $E_{electric} = V_{electric} \div d = 1500 \div (4.5 \times 10^{-3}) = 3.33... \times 10^5$ Vm⁻¹
= **3.3 × 10⁵ Vm⁻¹ (to 2 s.f.)** ***[1 mark]***

The field is perpendicular to the plates going from positive to negative. ***[1 mark]***

b) From the equation $E = \frac{V}{d}$, if the distance is doubled and the electric field strength remains constant, then the voltage must also double, so $V =$ **3000 V** ***[1 mark]***

4 Similarities — Any two from: Both gravitational and electric field strengths are forces per unit — gravitational field strength, g, is force per unit mass and electric field strength, $E_{electric}$, is force per unit positive charge. / Both gravitational and electric potentials are potential energy per unit — gravitational potential, V_{grav}, is potential energy per unit mass and electric potential, $V_{electric}$, is potential energy per unit positive charge. / Both are zero at infinity. / The force between two point masses is an inverse square law, and so is the force between two point charges. / The field lines for a point mass and the field lines for a negative point charge are the same. ***[2 marks — 1 mark for each correct similarity]***

Differences — Any one from: Gravitational forces are always attractive, whereas electric forces can be attractive or repulsive. / The size of an electric force depends on the medium between the charges, e.g. plastic or air. For gravitational forces, this makes no difference. / Objects can be shielded from electric fields, but not from gravitational fields. ***[1 mark for one correct difference]***

Page 57 — Millikan's Oil-Drop Experiment

1 a) The forces acting on the drop are its weight, acting downwards ***[1 mark]*** and the equally sized force due to the electric field, acting upwards ***[1 mark]***.

b) Electric force = weight, so $\frac{qV}{d} = mg$ and $q = \frac{mgd}{V}$ ***[1 mark]***

$q = \frac{1.63 \times 10^{-14} \times 9.81 \times 3.00 \times 10^{-2}}{5000} = 9.59... \times 10^{-19}$ C ***[1 mark]***

Divide by the magnitude of the charge on an electron:
$(9.59... \times 10^{-19}) \div (1.60 \times 10^{-19}) = 5.99... = 6$ (to 1 s.f.)
$\Rightarrow q =$ **6e** ***[1 mark]***

Answers

c) The forces on the oil drop as it falls are its weight and the viscous force from the air ***[1 mark]***. As the oil drop accelerates, the viscous force increases until it equals the oil drop's weight ***[1 mark]***. At this point, there is no resultant force on the oil drop, so it stops accelerating, but continues to fall at terminal velocity ***[1 mark]***.

d) At terminal velocity, $mg = 6\pi\eta rv$.

Rearranging, $v = \frac{mg}{6\pi\eta r}$ ***[1 mark]***

Find the radius of the oil drop, using mass = volume × density:

$m = \frac{4}{3}\pi r^3\rho$. So, $r^3 = \frac{3m}{4\pi\rho} = \frac{3\times1.63\times10^{-14}}{4\pi\times880} = 4.42... \times 10^{-18}$

and $r = 1.64... \times 10^{-6}$ m ***[1 mark]***.

So, $v = \frac{1.63\times10^{-14}\times9.81}{6\pi\times1.84\times10^{-5}\times1.64...\times10^{-6}} = 2.808... \times 10^{-4}$

$= \mathbf{2.8 \times 10^{-4}\ ms^{-1}}$ **(to 2 s.f.)** ***[1 mark]***

Page 59 — Charged Particles in Magnetic Fields

1 a) $F = qvB = 1.60 \times 10^{-19} \times 5.00 \times 10^6 \times 0.770$ ***[1 mark]***

$= \mathbf{6.16 \times 10^{-13}}$ **N** ***[1 mark]***

b) The force on the particle due to the electric field is $F = Eq$

If the particle travels in a straight line, the force due to the electric field must balance the force due to the magnetic field.

So equating, $Eq = qvB = 6.16 \times 10^{-13}$ N,

so $E = \frac{6.16\times10^{-13}}{1.60\times10^{-19}}$ ***[1 mark]***

$= \mathbf{3.85 \times 10^6\ NC^{-1}}$ **(to 3 s.f.)** ***[1 mark]***

2 a) The electron is moving in a circular path, so electromagnetic force = centripetal force

$qvB = \frac{mv^2}{r}$ ***[1 mark]***

Cancelling v from both sides gives: $qB = \frac{mv}{r}$ ***[1 mark]***

b) $r = \frac{mv}{qB} = \frac{(9.11\times10^{-31})\times(2.3\times10^7)}{(1.60\times10^{-19})\times(0.6\times10^{-3})} = 0.218...$

$= \mathbf{0.22\ m}$ ***[1 mark]***

Module 6: Section 3 — Probing Deep into Matter

Page 61 — Scattering to Determine Structure

1 a) Alpha particles and atomic nuclei are both positively charged ***[1 mark]***. If an alpha particle travels close to a nucleus, there will be a significant electrostatic force of repulsion between them ***[1 mark]***. This force deflects the alpha particle from its original path ***[1 mark]***.

b) The majority of the alpha particles are not scattered because the nucleus is a very small part of the whole atom and so most alpha particles don't get close enough to the nucleus to be scattered ***[1 mark]***.

2 Initial particle energy = 4.0 MeV

$4.0 \times 10^6 \times 1.60 \times 10^{-19} = 6.40 \times 10^{-13}$ J ***[1 mark]***

Electric potential energy $= \frac{kq_{proton}Q_{nucleus}}{r} = \frac{q_{proton}Q_{nucleus}}{4\pi\varepsilon_0 r}$

$= 6.40 \times 10^{-13}$ J at distance of closest approach ***[1 mark]***

$r = \frac{q_{proton}Q_{nucleus}}{4\pi\varepsilon_0\times6.40\times10^{-13}}$ ***[1 mark]***

$= \frac{(1\times1.60\times10^{-19})\times(13\times1.60\times10^{-19})}{4\pi\times(8.85\times10^{-12})\times(6.40\times10^{-13})}$

$= 4.675... \times 10^{-15} = \mathbf{4.7 \times 10^{-15}}$ **(to 2 s.f.)** ***[1 mark]***

Page 63 — Particles and Antiparticles

1 A proton, an electron and an antineutrino ***[1 mark]***. The electron and the antineutrino are leptons ***[1 mark]***. Leptons are not affected by the strong nuclear interaction, so the decay can't be due to the strong nuclear interaction ***[1 mark]***.

2 A neutrino / an antineutrino ***[1 mark]***

3 a) Antineutrons and antiprotons have a lepton number of 0, and positrons and antineutrinos both have a lepton number of –1 ***[1 mark]***.

$0 \neq 0 - 1 - 1$, so lepton number is not conserved ***[1 mark]***.

b) Antineutrons and antineutrinos both have a relative charge of 0, and antiprotons and electrons both have a relative charge of –1 ***[1 mark]***. $0 \neq -1 - 1 + 0$, so charge is not conserved ***[1 mark]***.

Page 65 — Pair Production and Annihilation

1 The creation of a particle of matter requires the creation of its antiparticle. In this case no antineutron has been produced ***[1 mark]***.

2 $e^+ + e^- \rightarrow \gamma + \gamma$ ***[1 mark]***. This is called annihilation ***[1 mark]***.

3 The energy of each particle is equal to $E_{rest} = m_pc^2$ (since kinetic energy is negligible).

When the proton and the antiproton annihilate, two identical photons are produced. So $2E_{photon} = 2m_pc^2$ and so $E_{photon} = m_pc^2$.

$E_{photon} = hf$, so $m_pc^2 = hf$,

$f = \frac{m_pc^2}{h}$ ***[1 mark]***

$= \frac{(1.673 \times 10^{-27}) \times (3.00 \times 10^8)^2}{(6.63 \times 10^{-34})}$ ***[1 mark]***

$= 2.271... \times 10^{23} = \mathbf{2.27 \times 10^{23}}$ **Hz (to 3 s.f.)** ***[1 mark]***

Page 67 — Quarks

1 udd ***[1 mark]***

2 proton = uud ***[1 mark]***

Relative charge of down quark = –1/3

Relative charge of up quark = 2/3

Total relative charge = 2/3 + 2/3 – 1/3 = +1 ***[1 mark]***

Page 69 — Particle Accelerators

1 $E_{rest} = m_pc^2 = 1.673 \times 10^{-27} \times (3.00 \times 10^8)^2$

$= 1.5057 \times 10^{-10}$ J ***[1 mark]***

$(1.5057 \times 10^{-10}) \div (1.60 \times 10^{-19}) = 9.410... \times 10^8$ eV ***[1 mark]***

$E_{total} = \gamma E_{rest}$ so $\gamma = E_{total}/E_{rest} = 500 \times 10^9 \div 9.410... \times 10^8$

$= 531.31... =$ **500 (to 1 s.f.)** ***[1 mark]***

2 $F = qvB$ so $B = F/qv$

$= (4.91 \times 10^{-13}) \div ((1.60 \times 10^{-19}) \times (6.82 \times 10^5))$

$= 4.499... =$ **4.50 T (to 3 s.f.)** ***[1 mark]***

Page 71 — Electron Energy Levels

1 The electron will escape the nucleus / not remain bound ***[1 mark]*** because its energy is greater than zero ***[1 mark]***.

2 a) $E = hf = 6.63 \times 10^{-34} \times 4.57 \times 10^{14} = 3.02991 \times 10^{-19}$

$= \mathbf{3.02 \times 10^{-19}}$ **J (to 3 s.f.)** ***[1 mark]***

b) $3.02991 \times 10^{-19} \div 1.60 \times 10^{-19} = 1.893...$ eV ***[1 mark]***

The difference between energy levels n = 3 and n = 2 is 3.4 – 1.5 = 1.9 eV. This is the closest figure to the energy of the photon, so the electron must have fallen between these energy levels ***[1 mark]***.

3 3.8×10^{-5} eV $= 3.8 \times 10^{-5} \times 1.60 \times 10^{-19}$

$= 6.08 \times 10^{-24}$ J ***[1 mark]***

$\Delta E = hf$ so $f = \Delta E \div h = (6.08 \times 10^{-24}) \div (6.63 \times 10^{-34})$ ***[1 mark]***

$= 9.170... \times 10^9 = \mathbf{9.2 \times 10^9}$ **Hz (to 2 s.f.)** ***[1 mark]***

Answers

Module 6: Section 4 — Ionising Radiation and Risk

Page 74 — Radioactive Emissions

1 ***5-6 marks:***

The answer gives a full description of an experiment to identify the type of radiation emitted by the source, including the results expected for alpha, beta and gamma emitters. The answer includes a discussion of correcting for background radiation. The answer has a clear and logical structure.

3-4 marks:

The answer describes an experiment to identify the type of radiation emitted by the source, including the results expected for alpha, beta and gamma emitters, but may omit some details. The answer has some structure.

1-2 marks:

There is some description of an experiment to identify the type of radiation emitted by the source, but the answer lacks detail. The answer has no clear structure.

0 marks:

No relevant information is given.

Here are some points your answer may include:

- Measure the background count over a sensible interval (e.g. 30 seconds) and divide by the time to get the count rate.
- Take at least three measurements and calculate an average background count rate.
- The background count rate should be subtracted from all of the results.
- Place the source in front of a Geiger-Müller tube attached to a Geiger counter, so that the counter records a high count rate when there is nothing between the source and tube.
- Insert different materials between the source and the tube, and record the count rate by measuring the count over a fixed time interval (selected depending on how active the source is) and dividing by this time.
- If the count rate drops significantly, then some of the radiation is being absorbed by the material. If it drops to zero after the background count rate has been subtracted, then all of the radiation is being absorbed.
- If the radiation is blocked by a piece of paper, the source is emitting alpha radiation.
- If the radiation is blocked by a thin (3 mm) sheet of aluminium, the source is a beta emitter.
- If a thick sheet of lead is needed to block the radiation, then the source is a gamma emitter.
- The radioactive source should only be handled using long-handled tongs and should not be pointed at anyone.
- The radioactive source should be kept in a lead-lined box when not in use.

2 a) $^{226}_{88}\text{Ra} \rightarrow ^{222}_{86}\text{Rn} + ^{4}_{2}\alpha$

[3 marks available — 1 mark for alpha particle, 1 mark each for proton and nucleon number of radon]

b) $^{40}_{19}\text{K} \rightarrow ^{40}_{20}\text{Ca} + ^{0}_{-1}\beta + ^{0}_{0}\bar{\nu}$

[3 marks available — 1 mark for beta particle and neutrino, 1 mark each for proton and nucleon number of calcium]

Page 77 — Dangers of Radiation

1 a) Any two of e.g.: They can cause cancerous tumours. / They can cause skin burns. / They can cause infertility. / They can cause hair loss. / They can cause radiation sickness. / They can cause death. ***[1 mark per correct answer, maximum 2 marks]***

b) Effective dose = absorbed dose × quality factor
For 0.6 Gy of alpha radiation, the effective dose is
0.6 × 20 = 12 Sv
For 9 Gy of beta radiation, the effective dose is
9 × 1 = 9 Sv ***[1 mark for both calculations]***
So, exposure to 0.6 Gy of alpha radiation would cause more damage to a sample of body tissue than 9 Gy of beta radiation ***[1 mark]***.

2 a) $T_{1/2} = \frac{\ln 2}{\lambda} = \frac{\ln 2}{3.21 \times 10^{-7}} = 2.159... \times 10^6$
$= \mathbf{2.16 \times 10^6}$ **s (to 3 s.f.)** ***[1 mark]***

b) energy deposited = activity × energy per decay × fraction of radiation absorbed
$= (5.03 \times 10^8) \times (2.9 \times 10^{-15}) \times 0.4$
$= 5.8348 \times 10^{-7}$ J ***[1 mark]***

absorbed dose = energy deposited ÷ mass
$= (5.8348 \times 10^{-7}) \div (4.2 \times 10^{-3})$
$= 1.38... \times 10^{-4} = \mathbf{1.4 \times 10^{-4}}$ **Gy (to 2 s.f.)** ***[1 mark]***

Remember to convert the mass to kg.

c) $0.50 \div (1.38... \times 10^{-4}) = 3599.09...$ = **3600 s (= 1 hour)** ***[1 mark]***

Page 79 — Binding Energy

1 a) There are 6 protons and 8 neutrons, so the mass of individual parts = (6 × 1.00728) + (8 × 1.00867) = 14.11304 u ***[1 mark]***
Mass of $^{14}_{6}$C nucleus = 13.99995 u
Mass defect = 13.99995 – 14.11304 = **–0.11309 u** ***[1 mark]***

b) mass in kg $= -0.11309 \times (1.661 \times 10^{-27})$
$= -1.8784... \times 10^{-28}$ kg ***[1 mark]***

binding energy, $E = mc^2 = (-1.8784... \times 10^{-28}) \times (3.00 \times 10^8)^2$
$= -1.69058... \times 10^{-11}$ J ***[1 mark]***

1 eV $= 1.60 \times 10^{-19}$ J,
so energy $= (-1.69058... \times 10^{-11}) \div (1.60 \times 10^{-19})$
$= -1.056... \times 10^8$ eV = **–106 MeV (to 3 s.f.)** ***[1 mark]***

2 a) Fusion ***[1 mark]***

b) There are two deuterium atoms before the reaction, each containing two nucleons, so:
binding energy before reaction = 2 × 2 × –1.12 = –4.48 MeV ***[1 mark]***
There is one helium atom after the reaction, containing three nucleons, and a free neutron with a binding energy of zero, so:
binding energy after reaction = (–2.57 × 3) + 0 = –7.71 MeV ***[1 mark]***
Energy released = difference in binding energy = –4.48 – (–7.71)
= **3.23 MeV** ***[1 mark]***

Page 81 — Nuclear Fission and Fusion

1 a) E.g. control rods limit the rate of fission by absorbing neutrons ***[1 mark]***. The number of neutrons absorbed by the rods is controlled by varying the amount they are inserted into the reactor ***[1 mark]***. A suitable material for the control rods is boron ***[1 mark]***.

b) In an emergency shut-down, the control rods are released into the reactor ***[1 mark]***. The control rods absorb the neutrons, and stop the reaction as quickly as possible ***[1 mark]***.

2 Advantages: e.g. the reaction in a nuclear reactor doesn't produce carbon dioxide ***[1 mark]***, it can produce a continuous supply of electricity, unlike some renewable sources ***[1 mark]***.
[1 mark for each advantage, maximum 2 marks]
Disadvantages — any two of: e.g. it could be dangerous if the reactor gets out of control, as a runaway reaction could cause an explosion ***[1 mark]*** / nuclear fission produces radioactive waste which is dangerous if it escapes into the environment ***[1 mark]*** / nuclear power-plants are expensive to build and decommission ***[1 mark]*** / nuclear waste has a long half-life, so has to be managed for a long time ***[1 mark]***.
[1 mark for each disadvantage, maximum 2 marks]

Index

Index

Index

PRBR62